Mastering Data Collection Tools and Instruments

Mastering Data Collection Tools and Instruments

A Simplified Guide

M. Rezaul Islam

PETER LANG
Chennai - Berlin - Bruxelles - Lausanne - New York - Oxford

Bibliographic information published by the Deutsche Nationalbibliothek.
The German National Library lists this publication in the German National Bibliography; detailed bibliographic data is available on the Internet at http://dnb.d-nb.de.

A catalogue record for this book is available from the British Library.

Library of Congress Cataloging-in-Publication Data

Names: Rezaul Islam, M., author.
Title: Mastering data collection tools and instruments : a simplified guide / M. Rezaul Islam.
Description: Oxford ; New York : Peter Lang, [2025] | Includes bibliographical references and index.
Identifiers: LCCN 2024051529 (print) | LCCN 2024051530 (ebook) | ISBN 9781803747187 (paperback) | ISBN 9781803747194 (ebook) | ISBN 9781803747200 (epub)
Subjects: LCSH: Sociology--Research.
Classification: LCC HM571 .R49 2025 (print) | LCC HM571 (ebook) | DDC 300.72–dc23/eng/20250130
LC record available at https://lccn.loc.gov/2024051529
LC ebook record available at https://lccn.loc.gov/2024051530

Cover design by Peter Lang Group AG

ISBN 978-1-80374-718-7 (print)
ISBN 978-1-80374-719-4 (ePDF)
ISBN 978-1-80374-720-0 (ePub)
DOI 10.3726/b22280

Published by Peter Lang Pvt Ltd, Chennai, India
info@peterlang.com - www.peterlang.com

This publication has been peer reviewed.

To my daughter Jemima Fairose and son Riasat Anjam,
Your unwavering love and boundless joy have filled my life with meaning and purpose. This book is dedicated to both of you, who inspire me every day to pursue knowledge and share it with the world.

Contents

Figures and Tables

Figures

Tables

Acknowledgments

I am especially grateful to my wife (Anjum Akter), daughter (Jemima Fairose), and son (Riasat Anjam) whose unwavering support and encouragement have been the driving force behind my endeavor to write this book, "Mastering Data Collection Tools and Instruments: A Simplified Guide." Their love and understanding have inspired me throughout this journey, and I am deeply thankful for their patience and belief in my work.

I also extend heartfelt thanks to my undergraduate, Masters, and Ph.D. students at the University of Dhaka, University of Malaya, Universiti Sains Malaysia (USM), Bangladesh University of Professionals (BUP), BRAC Institute of Governance and Development (BIGD) of BRAC University, University of Sharjah, University of Johannesburg, and the Bangladesh Civil Service Administration Academy. Your enthusiasm for learning and insightful discussions have significantly influenced the content of this book.

Furthermore, I appreciate the faculty and staff at numerous global and Bangladeshi universities where I have had the privilege to deliver over 100 lectures and public talks on different aspects of social research. Your engagement and feedback have been invaluable in shaping the ideas and concepts presented in this book.

Special thanks to Peter Lang and Ms. Indrani Dutta, Acquisitions Editor (India), for their guidance and support in bringing this manuscript to publication. Your expertise and dedication have been essential in making this book accessible to a broader audience.

Lastly, to my relatives and friends, thank you for your enduring support and belief in my aspirations. Your encouragement has been a constant source of strength and motivation.

Warm regards,
M. Rezaul Islam, Ph.D. (Nottingham)
Shaheed Giasuddin Ahmed Residential Area
Dhaka University Campus
Dhaka-1000, Bangladesh
E-mail: <rezauldu@gmail.com>, <rezaul.iswr@du.ac.bd>

Foreword

"Mastering Data Collection Tools and Instruments: A Simplified Guide" is a valuable asset for students, academics, and health professionals. It is also an essential resource for anyone involved in social research and offers a comprehensive yet accessible explanation of the tools and techniques necessary for effective data collection. The author, Professor M. Rezaul Islam, draws upon his wealth of experience and expertise to guide readers through the complexities of data collection in a simple to read manner that is evidence-based and practical.

The introductory chapters provide the reader with an insight into the context of data collection in social research and the pivotal role of selecting the right tools and instruments for the research objectives. From interviews and focus group discussions to the innovative use of technology, the author systematically covers the broad spectrum of data collection options, offering practical insights and real-world examples for their application.

Later chapters address the scaling and validating of instruments using piloting and statistical approaches. Understanding the ethical and diverse cultural context in data collection is critical and this is carefully considered in different demographic situations and across different research challenges. Finally, the book addresses the analysis and interpretation of the data and the need for continual skills development in collection and analysis.

"Mastering Data Collection Tools and Instruments: A Simplified Guide" is a critical foundation for all researchers for effective data collection and for the advancement of professional skills. The book is a highly welcome and timely text in the field of social research.

Glenn Laverack, Ph.D.,
Adjunct Full Professor at the College of Medicine & Health Sciences,
Institute of Public Health,
UAE University, Al Ain, United Arab Emirates.

Preface

Welcome to "Mastering Data Collection Tools and Instruments: A Simplified Guide." In an era driven by data, the art of collecting, analyzing, and interpreting information is paramount for informed decision-making and impactful research. This comprehensive guide is crafted to empower both novice and experienced researchers with the essential skills and knowledge needed to excel in data collection, a fundamental pillar of any data-driven endeavor.

What the Book Covers

"Mastering Data Collection Tools and Instruments: A Simplified Guide" is a comprehensive exploration of the multifaceted realm of data collection. This book is designed to equip a wide range of readers, from students and academics to professionals across diverse domains, with the essential knowledge and practical skills needed to excel in the art and science of data collection. The book commences with an in-depth examination of the Fundamentals of Data Collection, providing a strong foundation by elucidating key concepts, types of data, and a meticulous step-by-step guide to the data collection process. It then embarks on a journey through the intricate process of Selecting the Right Tools and Instruments, helping readers navigate the complexities of choosing the most appropriate data collection methods for their research endeavors, whether they be traditional or technologically advanced. In subsequent chapters, readers will delve into the diverse landscape of Data Collection Methods, ranging from interviews and observations to focus group discussions and beyond. A pivotal aspect of this guide is Designing Data Collection Instruments, where readers will learn how to craft meticulously tailored interview schedules, guidelines, and checklists to suit the unique demands of various research scenarios.

In the age of technology, the book doesn't shy away from the digital realm. It explores the integration of technology in data collection in Using Technology in Data Collection, offering insights into how modern tools can enhance efficiency and precision. Ensuring data's reliability and validity is a paramount concern addressed in Scaling and Validating Instruments, providing readers with techniques to maintain data integrity throughout the research process. Ethical considerations take center stage in Ethical Considerations in Data Collection, as readers navigate privacy, informed consent, and ethical dilemmas with wisdom and sensitivity.

Recognizing the richness of human diversity, Data Collection in Diverse Contexts offers invaluable guidance on conducting research across cultural, demographic, and geographical spectrums. In Analyzing and Interpreting Collected Data, the book demystifies the complexities of data cleaning, quantitative and qualitative analysis, and the art of deriving actionable insights from data. The journey concludes with a look at how to continually hone one's data collection skills, learn from experiences, and ultimately become an expert in the field of Enhancing Data Collection Skills.

Readers of the Book

This book's importance cannot be overstated, as it empowers readers to harness the transformative potential of data collection, be it for academic pursuits, professional decision-making, or bringing about meaningful change. Whether you are a newcomer or a seasoned data aficionado, this guide promises to elevate your proficiency in data collection and enhance your ability to navigate the intricate tapestry of research and analysis.

"Mastering Data Collection Tools and Instruments: A Simplified Guide" is tailored to cater to a diverse readership, recognizing that effective data collection is a fundamental skillset applicable across various domains. The book is crafted to be an invaluable resource for students and researchers, providing foundational knowledge and advanced insights to elevate their data collection capabilities, whether they are undergraduates embarking on their first research project or seasoned academic researchers.

Professionals in fields such as healthcare, business, social work, and education will also find this book instrumental in enhancing their data collection skills for evidence-based decision-making, program evaluation, and quality improvement. For data analysts and data scientists, the book offers insights into data quality, preparation, and interpretation, ensuring a solid foundation for more advanced analytics.

Policy makers and practitioners will discover essential insights into data-driven decision-making, enabling them to formulate effective policies and programs based on sound data-collection practices. Moreover, professionals in non-governmental organizations (NGOs) and social work can leverage the book's ethical considerations and cultural sensitivity insights to ensure data collection in diverse and sensitive contexts.

Consultants and market researchers engaged in consulting or market research will find this guide invaluable for crafting data collection strategies that yield meaningful insights and inform client recommendations. Finally, anyone interested in research, whether an inquisitive individual seeking to understand research methodologies or an enthusiast looking to embark on a personal research project, will find this book to be a clear and accessible entry point into the world of data collection.

In essence, "Mastering Data Collection Tools and Instruments" is a versatile guide designed to empower individuals from various walks of life with the knowledge and skills necessary to navigate the complex landscape of data collection effectively. It serves as an indispensable companion on the journey to becoming a proficient and ethical data collector.

How This Book Is Different?

"Mastering Data Collection Tools and Instruments: A Simplified Guide" distinguishes itself through a unique approach characterized by simplicity, clarity, and practicality. It stands out in the following ways:

This book places a strong emphasis on language simplicity, making it highly accessible to readers, including non-native English speakers. Complex ideas are presented with straightforward explanations, ensuring that individuals from diverse linguistic backgrounds can readily engage with the content. This inclusivity is a hallmark of the book's approach.

Unlike lengthy academic texts that delve into exhaustive detail, this guide offers a concise and focused reading experience. It distills information to its core components, delivering only the most critical details. Readers can benefit from a streamlined learning process, concentrating on the fundamental principles and practices of data collection without being inundated with excessive information.

One of the book's key strengths lies in its practical orientation. It serves as a hands-on handbook, providing readers with actionable insights and guidance. It's an invaluable resource for those seeking real-world assistance in data collection. By emphasizing essential concepts and practices, the book equips readers with the skills and knowledge needed to navigate the intricacies of data collection effectively.

This book stands out by limiting the use of references and citations. Instead of overwhelming readers with extensive reference lists, it adopts a more direct and reader-centric approach. The aim is to deliver the most up-to-date information without diverting the reader's focus. The book provides a concise, holistic view of data collection tools and instruments, ensuring that readers receive current and relevant insights.

In essence, "Mastering Data Collection Tools and Instruments: A Simplified Guide" prioritizes accessibility, practicality, and relevance. Its commitment to simplicity in language, conciseness in content, and applicability in practice makes it a valuable companion for anyone seeking to enhance their data collection skills, regardless of their linguistic background or level of expertise.

Importance of the Book

"Mastering Data Collection Tools and Instruments: A Simplified Guide" is a crucial resource in today's data-centric world, offering valuable insights into the art and science of effective data collection. This book holds significant importance across various dimensions. First and foremost, it empowers researchers, from beginners to seasoned professionals, by equipping them with essential skills for collecting high-quality data. In a world where research fuels progress, this guide contributes to credible research outcomes and advancements across diverse fields. Moreover, it plays a vital role in enhancing data-driven decision-making. Professionals across industries gain the expertise needed to harness data effectively, enabling them to base their actions on robust, reliable information.

Ethical considerations are at the heart of data collection, and this guide emphasizes their importance. It underscores ethical conduct, from respecting privacy to navigating complex ethical dilemmas, ensuring that data collection aligns with moral and legal standards. In a world as diverse as ours, cultural sensitivity is paramount. This book highlights the need to adapt data collection tools to different populations, fostering inclusivity and accuracy in research. Efficient resource utilization is a hallmark of effective data collection, and this guide helps organizations streamline their processes. It offers insights into selecting the right tools, technologies, and methods, thereby reducing wastage and redundancy.

As an educational resource, this book empowers students at all levels, nurturing a new generation of inquisitive minds capable of contributing to academia, innovation, and societal progress. Data collection transcends disciplines, and this guide bridges knowledge gaps between fields, making it accessible and relevant to individuals in academia, healthcare, business, social services, and beyond.

Finally, it places a strong emphasis on research quality assurance. High-quality data forms the foundation of reliable research, and this guide promotes well-designed instruments, rigorous validation, and meticulous analysis, ensuring that studies make meaningful contributions to their respective fields. In essence, this book serves as a guiding light in the field of data collection, illuminating the path toward excellence in research, decision-making, and ethical practice. Its importance lies in its potential to empower individuals and organizations to harness the power of data for the betterment of society.

Chapter Outlines

In this foundational Chapter 1: Introduction, readers are introduced to the critical importance of effective data collection in research. The chapter sets the stage by explaining the pivotal role of data collection tools and instruments, offering insights into why they are essential for conducting meaningful research. It also provides an overview of the book's objectives and structure, guiding readers on what to expect in the following chapters.

Building on the introduction, Chapter 2: Fundamentals of Data Collection describes the core concepts of data collection. Readers gain a deep understanding of what data collection is, its fundamental goals, and the key distinctions between quantitative and qualitative data. A step-by-step exploration of the data collection process provides a strong foundation for the chapters to come.

Chapter 3: Selecting the Right Tools and Instruments serves as a guide for readers on the critical task of tool and instrument selection. It outlines key considerations, such as the research objectives, that influence the choice of data collection tools. Practical advice is offered on how to match specific tools to research goals, and common data collection tools are explored in detail.

Chapter 4: Data Collection Methods: An Overview, readers embark on an in-depth exploration of various data collection methods. The chapter introduces and dissects methods such as interviews, observations, focus group discussions, and more. It provides not only a comprehensive understanding of each method but also insights into when and how to effectively employ them in research.

Chapter 5: Designing Data Collection Instruments guides readers through the art of crafting data collection instruments. It emphasizes the critical importance of clarity, relevance, and neutrality in question and instrument design. The chapter also highlights the necessity of pretesting and ensuring cultural sensitivity, ensuring that instruments yield reliable and meaningful data.

With technology increasingly integrated into research, Chapter 6: Using Technology in Data Collection explores how technology can enhance data collection. Readers learn about the integration of online surveys, digital questionnaires, and mobile apps to streamline data collection processes, making them more efficient and effective.

Ensuring the reliability and validity of data collection instruments is the focus of Chapter 7: Scaling and Validating Instruments. It provides readers with insights into maintaining measurement accuracy, the significance of pilot testing, and various statistical approaches to validation, all of which contribute to the credibility of research outcomes.

Ethical considerations are paramount in data collection. Chapter 8: Ethical Considerations in Data Collection guides readers through the complexities of respecting privacy, obtaining informed consent, and protecting participant rights. It also addresses the navigation of ethical dilemmas, ensuring that research is conducted with the utmost integrity.

In an increasingly diverse world, understanding how to adapt data collection tools to different cultural, demographic, and cross-cultural contexts is essential. Chapter 9: Data Collection in Diverse Contexts offers insights into ensuring cultural sensitivity in instrument design, adapting tools for varied demographics, and addressing the challenges that may arise.

Chapter 10: Analyzing and Interpreting Collected Data focuses on the post-data collection phase. Readers discover the importance of data cleaning and preparation for analysis, explore different approaches to quantitative and qualitative analysis, and learn how to draw meaningful conclusions from the collected data, empowering them to make informed decisions.

Chapter 11: Enhancing Data Collection Skills emphasizes the value of continual improvement in data collection techniques. It encourages readers to learn from their data collection experiences and offers guidance on how to build expertise in the field, ensuring that their data collection skills are continually honed.

The final Chapter 12: Conclusion serves as a comprehensive recap of key concepts and takeaways from the book. It empowers readers with a sense of mastery in data collection, offering them the knowledge and skills needed to excel in this critical aspect of research

CHAPTER 1

Introduction

ABSTRACT
Effective data collection is the cornerstone of empirical research, enabling evidence-based decision-making and the advancement of knowledge across diverse fields. This guide explores the vital role of data collection, the array of tools and instruments available, and the ethical considerations that underpin this fundamental research process. By delving into the nuances of data collection, researchers are empowered to navigate the complexities of research projects with skill, integrity, and innovation.

Keywords: Data collection, research methodology, tools and instruments, ethical considerations, evidence-based decision-making

1.1 Understanding Social Research

The section on understanding social research provides a comprehensive overview of the nature, characteristics, types, objectives, and importance of social research. It defines research as a systematic inquiry aimed at discovering, interpreting, and refining knowledge, with social research focusing specifically on human behavior interactions, and societies (For details, please see Islam, 2022a). The characteristics of social research, including its systematic approach, empirical basis, objectivity, and interdisciplinary nature, are highlighted. The section also delineates various types of social research, such as qualitative, quantitative, and mixed-methods approaches, as well as basic, applied, and action research. Furthermore, it elucidates the objectives of social research, emphasizing its role in understanding social phenomena, theory development, problem-solving, policy formulation, social change, empowerment, and education.

Finally, the importance of social research is underscored, emphasizing its contribution to evidence-based decision-making, social justice promotion, community engagement, interdisciplinary understanding, and continuous learning. Table 1.1 provides a concise yet comprehensive summary of the key points related to understanding social research, including its definition, characteristics, types, objectives, and importance.

Table 1.1. Summary Table: Understanding Social Research

Aspect	Details
Definition of Research	Systematic investigation aimed at discovering, interpreting, and refining knowledge in various fields.
Definition of Social Research	Focuses on understanding human behavior, interactions, and societies; aims to uncover patterns and underlying factors shaping social phenomena.
Characteristics of Social Research	
- Systematic Approach	Organized, methodical process ensuring reliability and validity.
- Empirical Basis	Based on direct observation or experimentation with real-world data.
- Objectivity and Impartiality	Strives to minimize bias and maintain neutrality.
- Generalizability	Seeks to draw broader conclusions applicable to similar contexts or groups.
- Theory-Driven	Guided by theoretical frameworks and models.
- Ethical Considerations	Adheres to ethical principles, protecting participants' rights and well-being.
- Interdisciplinary Nature	Integrates methodologies and perspectives from various disciplines.
- Practical Application	Aims to inform policy, shape interventions, and address societal issues.

Table 1.1. Continued

Aspect	Details
Types of Social Research	
- Qualitative Research	
-- Ethnography	Immersive observation in social settings to understand culture and practices.
-- Interviews	Structured or unstructured discussions to explore perspectives and experiences.
-- Focus Groups	Group discussions on specific topics to gather diverse viewpoints.
-- Case Studies	In-depth examination of single entities to understand unique phenomena.
-- Content Analysis	Systematic analysis of textual or visual content for patterns and themes.
-- Grounded Theory	Developing theories based on iterative data collection and analysis.
- Quantitative Research	
-- Surveys	Collecting numerical data from large samples for statistical analysis.
-- Experiments	Manipulating variables to observe effects under controlled conditions.
-- Observational Studies	Systematic observation in natural settings to analyze behaviors
-- Longitudinal Studies	Following subjects over time to track changes and trends.
-- Cross-sectional Studies	Comparing different groups at a single point in time.
-- Secondary Data Analysis	Analyzing pre-existing datasets for new insights.
- Mixed-Methods Research	
-- Triangulation	Using multiple data sources/methods to validate findings.

(*continued*)

Table 1.1. Continued

Aspect	Details
-- Sequential Design	Conducting qualitative and quantitative research in sequence.
-- Convergent Design	Simultaneous collection and analysis of qualitative and quantitative data.
Types of Social Research	
- Basic Research	
-- Aim	Expands knowledge without immediate practical application.
-- Focus	Theoretical exploration of social phenomena.
-- Example	Studying social media's impact on relationships theoretically.
- Applied Research	
-- Aim	Addresses specific practical problems with solutions.
-- Focus	Practical application to solve real-world issues.
-- Example	Evaluating the effectiveness of homelessness intervention programs.
- Action Research	
-- Aim	Combines research and action for social change.
-- Focus	Collaborative problem-solving and empowerment.
-- Example	Co-designing healthcare access strategies with communities.
Objectives of Social Research	
- Understanding Social Phenomena	Deepening understanding of human behavior and societies.
- Theory Development and Testing	Formulating and testing hypotheses within theoretical frameworks.

Table 1.1. Continued

Aspect	Details
- Exploration of Social Issues	Analyzing social problems to identify causes and solutions.
- Policy Formulation and Evaluation	Providing evidence for policy-making and assessing existing policies.
- Social Change and Advocacy	Promoting social justice and policy reforms.
- Empowerment and Community Development	Engaging communities in participatory research and development.
- Knowledge Transfer and Education	Disseminating research findings for academic and public benefit.
Importance of Social Research	
- Evidence-Based Decision Making	Providing data for informed policy and practice decisions.
- Social Problem Solving	Identifying and addressing social challenges.
- Promotion of Social Justice	Advocating for equity and human rights.
- Community Engagement	Facilitating community participation in research and solutions.
- Interdisciplinary Understanding	Fostering collaboration across fields for comprehensive insights.
- Cultural and Societal Enrichment	Documenting and preserving cultural practices.
- Continuous Learning	Encouraging ongoing learning and improvement through inquiry.

Source: Author

Research is a systematic investigation aimed at discovering, interpreting, and/or refining knowledge in a particular field of study. It involves the careful collection and analysis of data to answer questions, test hypotheses, or solve problems. Research is conducted across various disciplines and industries, including but not limited to science, technology, social sciences, humanities, and business.

Social research specifically focuses on understanding human behavior, interactions, and societies (Abusaleh & Anwar, 2022). It encompasses a wide range of topics such as social structures, cultural norms, individual and group behaviors, economic systems, political dynamics, and more (Bryman, 2016). Social research aims to uncover patterns, trends, and underlying factors that shape social phenomena.

Social research possesses several distinct characteristics that distinguish it from other forms of inquiry. These characteristics are fundamental to understanding the nature and purpose of social research. Here are some key characteristics:

Systematic Approach: Social research follows a systematic and organized process. It involves careful planning, execution, and analysis of data using established methodologies and techniques. This systematic approach ensures the reliability and validity of research findings.

Empirical Basis: Social research is grounded in empirical evidence obtained through direct observation or experimentation. It relies on data collected from real-world phenomena, such as surveys, interviews, observations, or experiments, rather than relying solely on theoretical speculation or anecdotal evidence.

Objectivity and Impartiality: Social researchers strive for objectivity and impartiality in their investigations. They aim to minimize bias and personal influence throughout the research process, from data collection to interpretation of results. This commitment to objectivity enhances the credibility and reliability of research findings.

Generalizability: Social research seeks to generate findings that can be generalized beyond the specific individuals or cases studied. While social researchers may focus on particular populations or contexts, they aim to draw broader conclusions that apply to similar groups or situations. The extent of generalizability depends on factors such as sample representativeness and research design (for detail, please see Creswell & Creswell, 2017; Creswell & Poth, 2016).

Theory-Driven: Social research is often theory-driven, meaning it is guided by established theoretical frameworks or conceptual models. Theoretical perspectives provide researchers with a foundation for formulating research questions, hypotheses, and analytical frameworks. By grounding their work in theory, researchers can contextualize their findings within existing knowledge and contribute to theoretical development.

Ethical Considerations: Ethical considerations are integral to social research, particularly when it involves human participants. Researchers must adhere to ethical principles and guidelines to ensure the protection of participants' rights, privacy, and well-being. This includes obtaining informed consent, maintaining confidentiality, and minimizing potential harm or risks.

Interdisciplinary Nature: Social research often adopts an interdisciplinary approach, drawing upon insights and methodologies from various disciplines such as sociology, psychology, anthropology, economics, and political science. This interdisciplinary perspective enriches the depth and breadth of understanding of complex social phenomena and facilitates holistic analyzes.

Practical Application: Social research is conducted with practical applications in mind. Its findings can inform policy-making, shape interventions, guide organizational decision-making, or contribute to academic scholarship. By addressing real-world issues and challenges, social research has the potential to effect positive change and improve societal well-being.

Overall, these characteristics collectively define social research as a systematic, empirical, objective, theory-driven, and ethically grounded inquiry aimed at advancing knowledge about human societies and addressing social issues.

Social research encompasses various types or methodologies that researchers employ to investigate different aspects of human behavior, interactions, and societies. These types of social research can be broadly categorized into qualitative and quantitative approaches (for details, please see Neuman & Robson, 2018), each with its own distinct methods and techniques. Here are the main types of social research:

Qualitative Research:

- Ethnography: Involves immersive observation and participation in a particular social setting or group to understand its culture, norms, and practices.
- Interviews: Conducting structured, semi-structured, or unstructured interviews with individuals or groups to explore their perspectives, experiences, and beliefs.

- Focus Groups: Bringing together a small group of participants to discuss specific topics or issues in-depth, allowing researchers to explore diverse viewpoints and interactions.
- Case Studies: In-depth examination of a single individual, group, or organization to understand unique circumstances, processes, or phenomena.
- Content Analysis: Systematic analysis of textual, visual, or audio-visual materials to identify patterns, themes, or discourses.
- Grounded Theory: Iterative process of data collection and analysis to develop theories or conceptual frameworks grounded in empirical evidence.

Quantitative Research:

- Surveys: Administering structured questionnaires or surveys to collect numerical data from a large sample of respondents, enabling statistical analysis and generalization of findings.
- Experiments: Manipulating variables and observing their effects on outcomes under controlled conditions to establish cause-and-effect relationships.
- Observational Studies: Systematic observation of behavior in natural or controlled settings to describe and analyze patterns, relationships, or phenomena.
- Longitudinal Studies: Following individuals or groups over an extended period to track changes, development, or trends over time.
- Cross-sectional Studies: Examining a population at a single point in time to compare different groups or variables.
- Secondary Data Analysis: Analyzing existing datasets, such as census data, survey archives, or administrative records, to explore research questions or test hypotheses.

Mixed-Methods Research:

- Triangulation: Integrating multiple data sources, methods, or perspectives to validate findings, enhance understanding, or address research questions more comprehensively.
- Sequential Design: Conducting qualitative and quantitative phases of research sequentially, with one method informing the other in a predetermined sequence.
- Convergent Design: Collecting and analyzing qualitative and quantitative data simultaneously to complement each other and provide a more holistic understanding of the research topic.

These types of social research offer researchers flexibility in choosing methods that best suit their research questions, objectives, and contexts, allowing for nuanced exploration and analysis of social phenomena.

Social research can be broadly categorized into three main types: basic research, applied research, and action research. Each type serves different purposes and has distinct characteristics:

Basic Research:

- Aim: Basic research, also known as pure or fundamental research, is conducted to expand knowledge and understanding of a particular phenomenon without any immediate practical application in mind.
- Focus: It focuses on theoretical development, exploring underlying principles, patterns, and relationships in social phenomena.
- Example: Investigating the impact of social media on interpersonal relationships without directly aiming to develop interventions or solutions.
- Characteristics: Basic research often involves exploring abstract concepts, testing hypotheses, and contributing to theoretical frameworks in social sciences.

Applied Research:

- Aim: Applied research aims to address specific practical problems, issues, or questions in real-world settings. Its primary goal is to generate solutions, inform decision-making, or improve practices.
- Focus: It focuses on the practical application of knowledge to solve concrete problems or meet specific needs.
- Example: Assessing the effectiveness of a community-based intervention program aimed at reducing homelessness.
- Characteristics: Applied research is typically solution-oriented, context-specific, and directly relevant to stakeholders, policymakers, or practitioners. It often involves collaboration between researchers and end-users to ensure the relevance and feasibility of findings.

Action Research:

- Aim: Action research is a participatory approach that combines research and action to address issues, empower communities, and bring about social change. It emphasizes collaboration, reflection, and continuous improvement.
- Focus: It focuses on identifying and implementing practical solutions to social problems through iterative cycles of planning, action, observation, and reflection.
- Example: Working with a community to co-design and implement strategies to improve access to healthcare services.
- Characteristics: Action research is characterized by active engagement with stakeholders, participatory methods, and a commitment to social justice. It often involves cycles of action and reflection, allowing for ongoing learning and adaptation.

Overall, these types of social research serve different purposes and have unique approaches, but they can also complement each other. Basic research lays the theoretical groundwork, applied research translates knowledge into practice, and action research engages communities in collaborative problem-solving and social change efforts.

Social research serves various objectives and holds significant importance in understanding, analyzing, and addressing complex social phenomena. Here's a detailed exploration of the objectives and importance of social research:

Objectives of Social Research:

Understanding Social Phenomena: One of the primary objectives of social research is to deepen our understanding of human behavior, interactions, institutions, and societies. By systematically investigating social phenomena, researchers aim to uncover underlying patterns, trends, and dynamics.

Theory Development and Testing: Social research contributes to the development and refinement of theories and conceptual frameworks in the social sciences. Researchers formulate hypotheses based on existing theories or empirical observations and test them through systematic inquiry.

Exploration of Social Issues and Problems: Social research aims to identify, analyze, and address social issues, problems, and challenges facing individuals, communities, and societies. It sheds light on the causes, consequences, and potential solutions to various social problems, such as poverty, inequality, discrimination, and environmental degradation.

Policy Formulation and Evaluation: Social research provides empirical evidence and insights that inform policy-making processes at local, national, and international levels. Policymakers rely on research findings to develop evidence-based policies, programs, and interventions aimed at addressing societal needs and promoting social well-being. Additionally, social research evaluates the effectiveness and impact of existing policies and programs to guide decision-making and improve outcomes.

Social Change and Advocacy: Social research plays a vital role in fostering social change, activism, and advocacy efforts. By documenting injustices, inequalities, and human rights violations, research raises awareness, mobilizes public support, and advocates for policy reforms and social justice initiatives.

Empowerment and Community Development: Social research empowers individuals, communities, and marginalized groups by giving them a voice, agency, and platform to articulate their concerns, experiences, and aspirations. Participatory research approaches, such as action research and community-based research, actively involve stakeholders in problem-solving and decision-making processes, leading to community development and empowerment (Islam, 2020, 2022c).

Knowledge Transfer and Education: Social research contributes to knowledge transfer and dissemination through academic publications, conferences, workshops, and other forms of communication. It enriches educational curricula, fosters critical thinking skills, and promotes lifelong learning in academic and non-academic settings.

Importance of Social Research:

Evidence-Based Decision Making: Social research provides policymakers, practitioners, and stakeholders with empirical evidence and insights to make informed decisions, allocate resources effectively, and address societal challenges.

Social Problem Solving: By identifying root causes, drivers, and consequences of social problems, social research informs strategies and interventions aimed at mitigating risks, promoting social justice, and improving quality of life.

Promotion of Social Justice and Equity: Social research contributes to the advancement of social justice, equity, and human rights by uncovering inequalities, discrimination, and injustices and advocating for policy reforms and interventions to address them.

Community Engagement and Participation: Social research engages communities, stakeholders, and marginalized groups in collaborative problem-solving, decision-making, and advocacy efforts, fostering a sense of ownership, empowerment, and solidarity.

Interdisciplinary Understanding: Social research fosters interdisciplinary collaboration and exchange of knowledge across diverse fields, enriching our understanding of complex social phenomena and promoting holistic approaches to social issues.

Cultural and Societal Enrichment: Social research documents and preserves cultural heritage, traditions, and practices, contributing to the enrichment and preservation of cultural diversity and societal cohesion.

Continuous Learning and Improvement: Social research encourages continuous learning, reflection, and adaptation by fostering a culture of inquiry, innovation, and evidence-based practice in academia, policymaking, and professional fields.

In summary, social research serves multifaceted objectives and holds profound importance in advancing knowledge, promoting social justice, fostering community development, and addressing pressing societal challenges. Its rigorous methodologies, empirical evidence, and interdisciplinary perspectives contribute to informed decision-making, social change, and the enhancement of human well-being.

1.2 Understanding Data Collection

At its core, data collection is a purposeful and systematic approach to gathering information from multiple sources, all to eventually analyze, interpret, and draw meaningful conclusions. This process is fundamental to research, decision-making, and problem-solving across numerous fields and industries. Data collection, in its various forms, involves a broad range of techniques, methods, and tools that enable researchers, professionals, and individuals to acquire raw data. This raw data serves as the foundation for further analysis and exploration. The sources of this data can be diverse, encompassing a multitude of information types.

These data types can span from numerical measurements and statistics, such as surveys that collect responses with quantitative values, to more qualitative descriptions and observations. For instance, qualitative data collection might involve in-depth interviews to capture detailed narratives or ethnographic fieldwork that delves into the nuances of a particular cultural context. What makes data collection truly powerful is its ability to provide a multifaceted perspective on a subject or problem. By gathering both quantitative and qualitative data, researchers can obtain a comprehensive understanding of the phenomenon under investigation. This multifaceted perspective is like having different lenses through which to view and comprehend a complex issue, enriching the research process and allowing for a more holistic interpretation of the data.

In essence, data collection serves as the initial step in the journey toward generating insights and knowledge. It is the process of gathering the building blocks from which patterns, trends, and meanings can be extracted. Moreover, the data collection process is highly adaptable, allowing researchers to tailor their methods to the unique demands of their research questions and the characteristics of the data sources. Whether in the realms of academia, business, healthcare, or social services, understanding and mastering data collection is a foundational skill in the modern world.

1.3 The Importance of Effective Data Collection

Data collection serves as the bedrock upon which research is built. It involves systematically gathering information to answer research questions, validate hypotheses, and inform decisions (Mabee et al., 1994). In today's multifaceted landscape, data collection serves as the cornerstone of informed decision-making across scientific research, business analysis, policy development, and beyond. It enables stakeholders to navigate complexities, identify trends, and make strategic choices with confidence. Table 1.2 provides a concise overview of the critical role data collection plays in advancing knowledge, validating hypotheses, assessing impact, identifying trends, and driving innovation across various fields.

Table 1.2. The Importance of Effective Data Collection

Topic	Summary
Informed Decision-Making	Data collection is crucial across various fields like scientific research, business analysis, and policy development, enabling stakeholders to make well-informed decisions by identifying trends, patterns, and correlations. It facilitates strategic choices, resource allocation, risk minimization, and the maximization of success opportunities.
Validating Research Hypotheses	Proper data collection aids in objectively testing and validating research hypotheses, transforming conjectures into supported or refuted propositions. It involves systematic gathering of data to evaluate hypotheses, refine theories, and drive evidence-based conclusions, contributing to the reliability of research findings and the advancement of knowledge.
Advancing Knowledge	Data collection expands human knowledge by enabling exploration, uncovering hidden relationships, and contributing to existing knowledge domains. It allows researchers to delve into uncharted territories, identify patterns, and bridge the gap between the known and unknown, facilitating the evolution of scientific understanding and the cumulative process of knowledge accumulation.
Assessing Impact	Data collection is essential for evaluating the impact of interventions, programs, and policies in sectors like public health, education, and social sciences. It helps stakeholders determine the effectiveness of efforts, measure objectives achievement, and make necessary adjustments to improve outcomes.
Identifying Trends and Patterns	Systematic analysis of data reveals trends, patterns, and anomalies crucial for predictive analysis and strategic planning across various sectors. It involves unearthing insights from raw data through careful collection and analysis, enabling stakeholders to anticipate changes, adapt strategies, and make informed decisions based on observed patterns and trends.

(*continued*)

Table 1.2. Continued

Topic	Summary
Evidence-Based Research	Reliable data collection forms the foundation of evidence-based research, grounding studies in real-world observations and measurements. In an era valuing informed decision-making, evidence-based research drives progress across disciplines by ensuring research integrity and validity through the systematic collection of reliable data.
Enhancing Accountability	Transparent and well-documented data collection processes foster accountability in research and decision-making by providing a clear pathway for others to review, replicate, or validate findings. When methods and data sources are traceable, and documentation is thorough, it enhances research integrity and strengthens the credibility of presented information, promoting accountability.
Mitigating Bias	Effective data collection methodologies aim to minimize bias and errors through thoughtful design and execution, ensuring collected data is representative and unbiased. By mitigating bias, researchers maintain the integrity of their findings and prevent distortions that could compromise the validity of conclusions drawn from the data.
Enabling Comparative Analysis	Data collected using standardized methods enables meaningful comparisons across different contexts, locations, or periods, allowing researchers to understand variations, identify trends, and make informed decisions based on broader perspectives. Standardized data collection facilitates comparative analysis, empowering researchers to draw insights that transcend individual instances and offer a broader understanding of the subject matter.
Driving Innovation	Data collection sparks innovation by revealing unexpected insights, prompting new hypotheses, and supporting continuous learning cycles. It serves as a compass guiding researchers through discovery journeys, propelling the world forward with fresh insights and groundbreaking innovations that result from the continuous process of systematically gathering and analyzing data.

Source: Author

Effective data collection is crucial for several key reasons:

1.3.1 Informed Decision-Making

Whether in scientific research, business analysis, policy development, or any other field, data collection provides the empirical evidence needed to make informed decisions. Accurate and reliable data allows stakeholders to identify trends, patterns, and correlations that guide strategic choices. In today's rapidly evolving and complex world, the ability to make well-informed decisions is paramount. Data collection serves as the bedrock upon which informed decision-making rests. Whether in scientific research, where groundbreaking discoveries are made, or in the realms of business, where strategies are crafted, and products are developed, the role of data collection cannot be overstated.

Scientists embark on journeys of exploration, formulating hypotheses and seeking answers to questions that advance human knowledge. Effective data collection underpins this process. By gathering empirical evidence, researchers can substantiate their hypotheses and refine their understanding of the phenomena they study. This enables them to arrive at conclusions that are based on rigorous observations rather than speculation.

In the realm of business, accurate data collection is essential for understanding markets, consumer behavior, and industry trends. Entrepreneurs, marketers, and strategists rely on data to gauge the demand for products and services, track consumer preferences, and identify emerging market opportunities. This data-driven approach empowers businesses to allocate resources effectively, minimize risks, and maximize their chances of success.

In policy development, data collection is indispensable for crafting informed and effective policies. Governments and organizations need reliable data to understand societal challenges, assess the impact of existing policies, and design interventions that address real-world issues. Policy decisions rooted in accurate data are more likely to yield positive outcomes and drive meaningful change.

The ability to identify trends, patterns, and correlations within collected data enhances decision-makers ability to navigate complexity. By analyzing data, stakeholders can identify subtle shifts in consumer behavior, economic indicators, public sentiment, and more. These insights allow them to anticipate changes, adapt strategies, and seize opportunities ahead of the curve.

Moreover, data collection is not merely about gathering information; it's about transforming raw data into actionable insights. Through careful analysis, stakeholders can extract meaning from data points and translate them into strategic choices. These choices range from adjusting manufacturing processes based on quality control data to shaping public health initiatives based on epidemiological trends.

In summary, data collection is a cornerstone of informed decision-making across diverse sectors. The power of accurate and reliable data lies in its ability to provide empirical evidence, illuminate patterns, and guide strategic choices. It empowers researchers, businesses, policymakers, and individuals alike to navigate complexity, mitigate risks, and harness opportunities for growth and progress. Through effective data collection, stakeholders gain the confidence to make decisions that shape the present and lay the foundation for a better future.

1.3.2 Validating Research Hypotheses

Research often begins with hypotheses that need to be validated or disproved (Ellis & Levy, 2009). Properly collected data helps researchers test these hypotheses objectively, leading to reliable conclusions and insights. At the heart of scientific inquiry lies the formulation of hypotheses—a process that drives discovery, innovation, and the expansion of knowledge. Hypotheses serve as educated guesses or propositions that researchers put forward to explain certain phenomena or relationships (Bhattacherjee, 2012). However, the strength of these hypotheses lies in their empirical validation, which is where proper data collection plays a crucial role.

Imagine a scenario where a biologist posits that a certain environmental factor influences the growth rate of a particular species of plants.

This hypothesis, while intriguing, remains speculative until subjected to rigorous testing through data collection. The process of systematically gathering data related to the growth rates of these plants in different environmental conditions transforms the hypothesis from a conjecture into a hypothesis that is either supported or refuted by evidence.

By collecting data in a controlled and structured manner, researchers can objectively evaluate their hypotheses. This data-driven approach is a hallmark of the scientific method—a systematic process of inquiry that involves formulating hypotheses, collecting, and analyzing data, and drawing conclusions. Properly collected data acts as the cornerstone of this method, allowing researchers to discern patterns, trends, and correlations that either affirm or challenge their initial assumptions.

The role of data in hypothesis validation extends beyond a simple binary outcome of "supported" or "rejected" (Braumoeller, 2004). It enables researchers to refine and modify their hypotheses based on empirical observations. If the collected data do not align with the initial hypothesis, researchers can use this information to revise their theories, consider alternative explanations, and refine their research questions. This iterative process of hypothesis testing and data collection drives the evolution of scientific understanding and ensures that conclusions are based on evidence rather than preconceived notions.

Importantly, properly collected data contributes to the reliability of research findings. When hypotheses are subjected to rigorous empirical validation, the resulting conclusions are more robust and credible (Abowitz & Toole, 2010). These findings are more likely to withstand scrutiny from peers, contribute to the advancement of knowledge, and inform subsequent research directions.

In summary, the process of hypothesis validation is a cornerstone of scientific inquiry, and data collection is the engine that drives this process. Properly collected data transforms conjectures into testable hypotheses and enables researchers to objectively assess their theories. By subjecting hypotheses to empirical scrutiny, researchers ensure that their conclusions are grounded in evidence, contributing to the growth of knowledge and the advancement of fields across the spectrum of human inquiry.

1.3.3 Advancing Knowledge

Data collection contributes to the expansion of human knowledge. It enables researchers to explore new frontiers, uncover hidden relationships, and contribute to the existing body of knowledge within their respective domains. Human progress is intricately linked to the accumulation of knowledge. From the earliest explorations of our world to the intricate scientific inquiries of today, the quest for understanding drives innovation and growth. Data collection stands as a cornerstone of this pursuit, enabling researchers to journey into uncharted territories, unveil hidden connections, and add building blocks to the vast edifice of human knowledge.

Picture a scientist embarking on a research journey to study a rare and enigmatic species in a remote ecosystem. Through meticulous data collection, they systematically document behavioral patterns, ecological interactions, and environmental factors. The data they gather shed light on previously unknown aspects of this species, illuminating its role in the ecosystem and potentially unveiling insights that can inform conservation efforts or inspire new scientific inquiries.

By collecting data with precision and rigor, researchers have the power to uncover hidden relationships that may have eluded previous generations (Sagor, 2000). They can identify patterns, trends, and correlations that transform fragmented information into coherent narratives. These narratives, in turn, contribute to the broader understanding of natural phenomena, societal dynamics, technological advancements, and countless other realms of human exploration.

Data collection serves as a bridge between the known and the unknown. It empowers researchers to explore new frontiers, challenge conventional wisdom, and expand the boundaries of human understanding (Cobb et al., 2013). Just as explorers once navigated uncharted seas and mapped unfamiliar landscapes, modern-day researchers use data collection to venture into unexplored realms of knowledge (Cohen et al., 2018).

Moreover, the data collected by one generation of researchers often becomes the foundation upon which future inquiries are built. Each observation, measurement, and insight add a thread to the rich tapestry of human knowledge. This cumulative process allows subsequent researchers to build on past findings, refine theories, and delve deeper into the intricacies of their respective fields.

Data collection is not limited to scientific research alone; it extends to virtually every facet of human endeavor (Sarantakos, 2017). In business, data collection drives market research, consumer behavior analysis, and strategic decision-making. In the social sciences, data collection sheds light on cultural dynamics, societal trends, and human behavior. In medicine, data collection informs diagnoses, treatment plans, and the development of life-saving interventions.

In summary, data collection is a cornerstone of human progress, enabling researchers to explore new horizons, uncover hidden relationships, and contribute to the ever-expanding realm of human knowledge. With each piece of data collected, humanity takes a step forward on the path of understanding, innovation, and enlightenment. Through rigorous data collection, we continue to push the boundaries of what is known, weaving a tapestry of discovery that enriches our understanding of the world and fuels the fires of curiosity that have driven us for centuries.

1.3.4 Assessing Impact

In various sectors such as public health, education, and social sciences, data collection is essential for evaluating the impact of interventions, programs, and policies. It allows stakeholders to determine whether objectives have been achieved and what adjustments may be necessary. In the dynamic landscape of public health, education, social sciences, and beyond, the effectiveness of interventions, programs, and policies is paramount. How do we know if a vaccination campaign is reducing disease prevalence? How can we measure the effectiveness of an educational initiative? How do we gauge the impact of social policies on vulnerable populations? The answer lies in data collection—a process that serves as a powerful lens through which the impact of these efforts can be assessed.

1.3.4.1 Monitoring and Evaluation

Data collection plays a pivotal role in the monitoring and evaluation of interventions, programs, and policies. Monitoring involves the systematic collection of data over time to track progress and ensure that activities are being carried out as planned. Evaluation, on the other hand, focuses on assessing the outcomes and impacts of these activities. Imagine a scenario where a public health agency implements a vaccination campaign to combat a prevalent disease. Regular data collection on vaccination rates, disease incidence, and other relevant metrics allows stakeholders to monitor the campaign's progress. By analyzing collected data, they can determine whether the campaign is reaching its intended target, whether coverage rates are sufficient, and whether any adjustments are needed to achieve the desired impact.

1.3.4.2 Evidence-Based Decision-Making

Data collection in these sectors empowers stakeholders to make evidence-based decisions. Instead of relying on assumptions or anecdotal evidence, decision-makers can draw on concrete data to assess whether interventions are achieving their objectives. This evidence-based approach informs resource allocation, policy adjustments, and strategic planning. For instance, in the field of education, data collection helps educators and policymakers understand the effectiveness of teaching methods, curriculum changes, and educational interventions. By analyzing student performance data, dropout rates, and other educational indicators, they can identify areas of improvement and tailor their approaches to better meet the needs of students.

1.3.4.3 Continuous Improvement

Data collection not only provides insights into impact but also enables continuous improvement. When data highlights areas where interventions are falling short of their objectives, stakeholders can adapt their strategies, make necessary adjustments, and refine their approaches. This iterative process ensures that resources are allocated effectively and that interventions evolve in response to real-world outcomes. Consider a social program aimed at reducing poverty rates in a specific community. Data collection can help assess whether the program is achieving its intended impact. If data reveals that the poverty rates have not significantly decreased, stakeholders can use this information to redesign the program, refine its implementation, or target it more effectively to achieve better results.

1.3.4.4 Evidence of Accountability

In addition to facilitating improvement, data collection establishes a foundation for accountability. When stakeholders are armed with data that demonstrates the impact of interventions, they can be held accountable to the communities they serve, donors, and other stakeholders. Transparency in reporting and using data to showcase progress or challenges bolsters trust and ensures that efforts remain focused on achieving meaningful outcomes. In conclusion, data collection is a linchpin for assessing the impact of interventions, programs, and policies in sectors ranging from public health to education and social sciences. Through systematic data collection, stakeholders can determine whether objectives have been met, identify areas for improvement, and make evidence-based decisions that lead to positive change. As the world continues to evolve, data collection stands as a powerful tool to drive progress, refine strategies, and ensure that efforts are effective and responsive to the needs of the communities they serve.

1.3.5 Identifying Trends and Patterns

Data, when analyzed systematically, reveals trends, patterns, and anomalies that might otherwise go unnoticed. These insights are crucial for predictive analysis and strategic planning. In the sea of information that our world generates, data serves as a treasure trove of insights waiting to be unearthed. However, this treasure is often hidden within the vast expanse of raw numbers and observations. It is through systematic data collection and thoughtful analysis that these hidden gems are revealed—trends that unfold over time, patterns that repeat themselves, and anomalies that disrupt the norm. These insights are not only fascinating from an academic perspective; they are also crucial for predictive analysis and strategic planning across a multitude of sectors.

1.3.5.1 Uncovering Trends

Trends are like the currents that flow beneath the surface of a river, guiding its course. Data collected over time allows researchers and analysts to observe the subtle shifts and changes that shape our world. For instance, in the realm of climate science, data on temperature variations and sea level rises help identify long-term trends that inform our understanding of climate change. Similarly, economic indicators such as employment rates, inflation, and GDP growth reveal economic trends that influence policies and business decisions.

1.3.5.2 Discerning Patterns

Patterns are the threads that weave together data points into coherent narratives. By systematically analyzing data, researchers can discern repeating sequences, relationships, and interactions that might not be immediately apparent. For instance, in healthcare, data analysis might reveal patterns in patient symptoms that lead to earlier and more accurate diagnoses. In cybersecurity, analyzing patterns of network activity can help detect unusual behaviors that signify potential threats.

1.3.5.3 Identifying Anomalies

Anomalies are the outliers that defy expectations and provide valuable insights. These outliers might be unexpected spikes in customer complaints, sudden shifts in market behavior, or anomalies in medical test results. While anomalies might seem like noise at first glance, they often signify significant events or changes that warrant further investigation. By identifying and understanding anomalies, stakeholders can uncover underlying causes, mitigate risks, and respond proactively.

1.3.5.4 Predictive Analysis and Strategic Planning

Identifying trends, discerning patterns, and understanding anomalies contribute to predictive analysis—a process that involves using historical data to forecast future events. Predictive analysis empowers stakeholders to anticipate trends, assess potential outcomes, and plan strategies that are well-informed and future-ready. Businesses can use predictive analysis to forecast consumer behavior and adapt marketing strategies accordingly. Public health agencies can predict disease outbreaks based on historical patterns of infection.

Strategic planning, in turn, hinges on these predictive insights. Armed with an understanding of trends and patterns, decision-makers can craft strategies that capitalize on opportunities and mitigate risks. This proactive approach to planning ensures that organizations are well-positioned to navigate changing landscapes and make informed choices.

In summary, data collection and analysis enable us to uncover the hidden stories within the numbers. Identifying trends, discerning patterns, and understanding anomalies are all crucial aspects of making sense of complex data sets. These insights not only deepen our understanding of the world around us but also empower us to make predictions, plan strategically, and make well-informed decisions that shape the future. As we continue to gather and analyze data, we unlock the potential to anticipate change, seize opportunities, and build a more informed and resilient world.

1.3.6 Evidence-Based Research

In an era where evidence-based practices are highly valued, reliable data collection provides the foundation for research that is grounded in real-world observations and measurements. The pursuit of knowledge has evolved from the realm of conjecture and speculation to a realm deeply rooted in evidence. In an era where informed decision-making and effective problem-solving are prized, evidence-based research has become the cornerstone of progress across disciplines. Reliable data collection serves as the bedrock upon which evidence-based practices are built, grounding research in real-world observations and measurements.

1.3.6.1 The Shift to Evidence-Based Practices

Evidence-based practices denote an approach where decisions and actions are guided by rigorous research findings, rather than tradition, intuition, or anecdotal evidence. This shift has been fueled by the realization that relying on empirical evidence provides a more reliable foundation for achieving desired outcomes. Consider the medical field. Evidence-based medicine has transformed how healthcare is delivered, with decisions about treatments and interventions guided by rigorous clinical trials and data-driven research. By collecting and analyzing data from large patient populations, medical practitioners can make informed choices that optimize patient care and improve health outcomes.

1.3.6.2 The Role of Reliable Data Collection

At the heart of evidence-based research lies reliable data collection. The data collected serves as the empirical evidence upon which conclusions are drawn, theories are formulated, and practices are refined. When data is collected with precision, rigor, and a commitment to accuracy, it ensures that research findings are robust and trustworthy. Reliable data collection involves carefully designing research protocols, selecting appropriate measurement tools, and systematically recording observations. It requires adherence to standardized methods that minimize bias and errors, allowing researchers to confidently rely on the data they gather.

1.3.6.3 Translating Research into Action

The value of evidence-based research lies not only in its academic significance but also in its practical applications (Biesta, 2007). The insights gained from rigorous data collection inform policy decisions, guide the development of interventions, and shape best practices in various fields. In education, evidence-based research informs teaching methodologies, curriculum design, and assessment strategies. By analyzing student performance data, educators can identify effective teaching techniques, learning interventions, and areas where improvement is needed. In environmental science, evidence-based research influences conservation policies, land management strategies, and sustainability initiatives. Through data collection and analysis, researchers can assess the impact of human activities on ecosystems, guide habitat restoration efforts, and advocate for policies that protect biodiversity.

1.3.6.4 Enhancing Accountability and Transparency

Evidence-based research enhances accountability and transparency. When research findings are derived from reliable data collection and thorough analysis, they can be shared, scrutinized, and replicated by other researchers. This transparency promotes the peer review process and ensures that knowledge is built upon a solid foundation.

In summary, the era of evidence-based practices is a testament to the power of rigorous data collection. As disciplines across the spectrum embrace the value of grounded research, the importance of reliable data collection becomes clear. Through meticulous and systematic data collection, researchers contribute to a knowledge ecosystem that is informed by real-world observations and measurements. This ecosystem, in turn, fuels progress, drives innovation and empowers decision-makers to make choices that are anchored in evidence and optimized for success.

1.3.7 Enhancing Accountability

Transparent and well-documented data collection processes enhance accountability in research and decision-making. When methods are clearly outlined and data sources are traceable, others can review, replicate, or validate the findings. Accountability serves as a cornerstone of integrity in research and decision-making. Transparent and well-documented data collection processes play a pivotal role in fostering accountability by providing a clear and verifiable pathway for others to assess and validate the information presented. When research methods are openly shared, data sources are traceable, and documentation is thorough, the foundations of accountability are solidified.

1.3.7.1 Openness and Transparency

Transparency is the keystone of accountability. Transparent data collection processes involve openly sharing how data was gathered, measured, and analyzed. This transparency enables other researchers, experts, and stakeholders to gain a comprehensive understanding of the methodology used, ensuring that there are no hidden or biased practices that could skew the results. For example, in climate science, transparent data collection practices involve disclosing the methods used to gather temperature, atmospheric, and oceanic data. By sharing these methods, the scientific community can assess the validity of the data and ensure that conclusions are not influenced by undisclosed biases.

1.3.7.2 Replicability and Validation

Transparent and well-documented data collection processes facilitate replicability—the ability of other researchers to repeat the study using the same methods and data sources. Replicability is a critical aspect of scientific inquiry, as it allows others to validate and verify research findings. When methods are clearly outlined and data sources are traceable, replication becomes feasible. Replicability acts as a powerful quality control mechanism. If other researchers can replicate a study's findings using the same methods and data, it lends credibility to the original research and reinforces the validity of the conclusions.

1.3.7.3 Peer Review and Quality Assurance

Transparent data collection processes facilitate the peer review process—an essential component of the scientific method. Peer review involves subjecting research to scrutiny by experts in the same field. Transparent documentation of data collection methods allows reviewers to assess the research's rigor, identify potential shortcomings, and offer insights for improvement. Additionally, transparent data collection contributes to quality assurance. Well-documented methods and traceable data sources enable organizations, institutions, and regulatory bodies to ensure that research adheres to established standards of integrity and accuracy.

1.3.7.4 Accountability in Decision-Making

Accountability is not limited to research; it extends to decision-making processes across various sectors. Transparent data collection supports accountable decision-making by providing decision-makers with a clear understanding of the evidence that informs their choices (Farley-Ripple, 2012). Whether in policy development, business strategies, or public initiatives, transparent data collection processes allow stakeholders to evaluate the information that underpins decisions. For instance, in policymaking, transparency in data collection ensures that decisions are based on accurate and unbiased information. Policymakers can assess the quality and reliability of the data used to formulate policies, fostering public trust and confidence in the decision-making process.

In summary, enhancing accountability through transparent and well-documented data collection processes is essential for maintaining the integrity of research and decision-making. By openly sharing methods, making data sources traceable, and facilitating replicability, researchers and decision-makers ensure that their work is subject to scrutiny, validation, and peer review. This commitment to transparency not only strengthens the foundations of knowledge but also upholds the principles of accountability that underpin responsible research and informed decision-making in a diverse range of fields.

1.3.8 Mitigating Bias

Effective data collection methodologies are designed to minimize bias and errors (Pannucci & Wilkins, 2010; Biemer, 2010). Thoughtful design and execution help ensure that collected data is representative and unbiased. In the pursuit of accurate and reliable data, the specter of bias looms large. Bias can distort data, compromise findings, and lead to erroneous conclusions. Effective data collection methodologies are meticulously crafted to minimize bias and errors, ensuring that the collected data is not only representative but also devoid of undue influence that could taint its integrity.

1.3.8.1 Understanding Bias

Bias refers to the systematic distortion of data that occurs when certain factors influence the collection, analysis, or interpretation of information. These factors can stem from a variety of sources, including the research process, the researcher's perspective, or the participants themselves. For example, selection bias might occur if participants are not chosen randomly, leading to a sample that does not accurately reflect the larger population. Confirmation bias might influence a researcher to only focus on data that supports their preconceived notions, ignoring information that contradicts their beliefs.

1.3.8.2 Minimizing Bias through Design

Effective data collection methodologies prioritize reducing bias from the outset. Researchers take meticulous care in designing their studies, considering factors that could introduce bias and planning strategies to mitigate them. Random sampling is a common technique used to minimize selection bias. By selecting participants randomly from the population of interest, researchers ensure that the sample is representative and not skewed towards a particular subset. Blinding, a technique often used in experiments, helps mitigate observer bias. In a double-blind study, neither the participants nor the researchers know which participants are receiving the treatment and which are receiving a placebo, reducing the potential for biased observations or interpretations.

1.3.8.3 Reducing Researcher Bias

Researchers' own biases can inadvertently influence data collection and analysis (Jootun et al., 2009). Effective methodologies acknowledge this potential and incorporate measures to mitigate researcher bias. Structured data collection protocols, such as standardized questionnaires, help ensure that participants are asked the same questions in the same way, reducing the likelihood of personal bias affecting the data. Peer review and collaboration also play a role in reducing researcher bias. When multiple researchers are involved in data collection and analysis, differing perspectives can act as checks and balances against individual biases.

1.3.8.4 Participant Bias and Ethical Considerations

Participants can introduce bias by providing inaccurate or skewed responses due to social desirability bias (responding in socially acceptable ways) or demand characteristics (changing behavior based on perceived expectations). To mitigate participant bias, researchers use techniques such as anonymous surveys to encourage honest responses. Ethical considerations are also paramount, ensuring that participants are fully informed about the research, their rights, and the potential implications of their participation.

1.3.8.5 Analyzing and Addressing Bias

Even with the most careful planning, bias can still emerge. Researchers must be vigilant in identifying and addressing bias during data analysis and interpretation. Sensitivity analyzes, subgroup analyzes, and exploring alternative explanations can help researchers assess the robustness of their findings.

In summary, mitigating bias is a central tenet of effective data collection methodologies. By thoughtfully designing research, minimizing researcher bias, and considering ethical considerations, researchers strive to ensure that the data they collect is representative, accurate, and free from undue influence. The commitment to mitigating bias upholds the integrity of research, enhances the validity of findings, and contributes to the advancement of knowledge that can be confidently relied upon by decision-makers across a range of fields.

1.3.9 Enabling Comparative Analysis

Data collected using standardized methods can be compared across different contexts, locations, or periods, enabling researchers and practitioners to draw meaningful comparisons and conclusions. The ability to draw meaningful comparisons is a linchpin of research and decision-making. It allows us to understand variations, identify trends, and make informed choices based on a broader context. Data collected using standardized methods serves as a powerful tool for enabling comparative analysis, empowering researchers and practitioners to extract insights that transcend individual instances and offer a broader perspective.

1.3.9.1 Standardization for Consistency

Standardized data collection methods involve using uniform procedures and protocols to gather information. This standardization ensures that data is collected consistently, regardless of the context, location, or period. When data is collected using the same methods, it becomes comparable, facilitating meaningful cross-comparisons. For example, imagine a global study on childhood literacy rates. If researchers in different countries use standardized assessment tools to measure literacy skills, the resulting data can be compared directly. This standardized approach allows for insights into global trends, regional disparities, and the impact of various interventions.

1.3.9.2 Comparing across Contexts

Comparative analysis enables researchers to compare data collected in different contexts to uncover similarities, differences, and factors that contribute to variations. By analyzing data collected using standardized methods, researchers can identify patterns that transcend geographical, cultural, or socio-economic boundaries. For instance, consider studies on air quality in urban areas across the world. By using standardized measurement techniques, researchers can compare air quality data from different cities. This enables them to identify common pollutants, assess the effectiveness of air quality policies, and share best practices to improve urban environments.

1.3.9.3 Identifying Trends and Changes

Comparative analysis also facilitates the identification of trends and changes over time. By collecting data using consistent methods at multiple time points, researchers can track shifts in behavior, outcomes, or phenomena. In the field of economics, for instance, standardized data collection on inflation rates over decades allows economists to analyze long-term trends and make predictions about future economic conditions. Similarly, in epidemiology, standardized data on disease prevalence can reveal patterns of outbreaks and help anticipate potential health crises.

1.3.9.4 Informed Decision-Making

Comparative analysis extends beyond research to inform decision-making in various sectors. Policymakers, business leaders, and educators can draw insights from comparisons to guide their choices. In international development, for example, comparing data on education, healthcare, and economic indicators among countries can help allocate resources effectively and design interventions that address specific challenges.

1.3.9.5 Addressing Challenges and Opportunities

Comparative analysis also empowers us to address challenges and seize opportunities. By analyzing data from different contexts, we can identify successful strategies, lessons learned, and potential areas for improvement. When faced with issues like climate change, comparative analysis of data on greenhouse gas emissions, deforestation rates, and energy consumption can guide international cooperation and policy development to mitigate global environmental challenges.

In summary, standardized data collection methods pave the way for comparative analysis that transcends individual cases and offers insights with broader implications. By enabling cross-context, cross-location, and cross-temporal comparisons, standardized data collection empowers researchers and decision-makers to understand variations, identify trends, and make evidence-based choices that drive progress and enhance our understanding of complex issues.

1.3.10 Driving Innovation

Data collection often reveals unexpected insights that can lead to innovations, hypotheses, and areas of exploration. It sparks curiosity and supports a continuous cycle of learning. The journey of discovery is often unpredictable, and data collection acts as a compass that guides us through uncharted territories. While data collection aims to provide answers, it also has the uncanny ability to uncover new questions, spark curiosity, and ignite innovation. Through the process of systematically gathering and analyzing data, researchers and practitioners embark on a continuous cycle of learning, propelling the world forward with fresh insights and groundbreaking innovations.

1.3.10.1 Uncovering the Unexpected

Data collection is like a treasure hunt where unexpected gems lie hidden within the data points. Researchers may start with specific questions in mind, but the act of collecting and examining data often unearths unexpected patterns, correlations, and anomalies. These surprises can challenge existing theories, prompt new hypotheses, and open up avenues for exploration that were previously unconsidered. For instance, in astronomy, data collected from telescopes may reveal anomalies in the light emitted from distant galaxies, leading to the discovery of previously unknown cosmic phenomena. In the world of business, customer behavior data may reveal unexpected trends that inspire innovative marketing strategies.

1.3.10.2 Fostering Curiosity and Creativity

Data collection fuels curiosity and creativity. As researchers dive into data sets, they become detectives, piecing together fragments to create a coherent picture. This process can spark curiosity about the underlying mechanisms, patterns, and causes behind the observed phenomena. When data collected from a rainforest ecosystem shows unexpected interactions between species, scientists might be inspired to investigate the ecological mechanisms that drive these interactions. The data collected might not only answer questions but also ignite a curiosity-driven journey of exploration.

1.3.10.3 Inspiring New Innovations

Innovation often arises from the collision of existing knowledge with fresh insights gained through data collection. These insights can illuminate gaps in current practices, suggesting new ways of solving problems or addressing challenges. Consider the field of medicine, where data collected from patient outcomes may reveal patterns of treatment effectiveness. These patterns can inspire the development of innovative therapies or medical devices that improve patient care.

1.3.10.4 Supporting a Cycle of Learning

Data collection is not a linear process; it's a cycle of learning that encourages iteration and refinement. Insights gained from data collection feed back into the research process, leading to the formulation of new questions, hypotheses, and research directions. For example, social scientists studying human behavior may collect data on societal trends, uncovering unexpected shifts in cultural norms. These findings can lead to the formulation of new research questions, further data collection, and the continuous evolution of understanding.

1.3.10.5 Filling Gaps in Knowledge

As data collection reveals gaps in our understanding, it also provides the motivation to fill those gaps. Researchers and practitioners are driven to collect more data, explore new methods, and innovate to address the questions that emerge from the initial data collection. For instance, a study on the effects of pollution on marine life may reveal gaps in data about the specific impacts on certain species. This gap may drive researchers to develop new methodologies for data collection underwater or inspire collaborative efforts to fill the knowledge void.

In summary, data collection is not just a means of finding answers; it's a catalyst for innovation, curiosity, and continuous learning. The insights uncovered through data collection often lead to unexpected revelations, new questions, and opportunities for exploration. This cycle of discovery and innovation pushes the boundaries of human understanding, propelling us toward solutions and advancements that shape the future.

In the subsequent chapters of this guide, we will explore the principles, methods, tools, and instruments necessary to master the art of data collection. From understanding the different data collection methods to selecting appropriate tools, designing effective instruments, and ensuring ethical considerations, this guide equips you with the knowledge to become a proficient data collector. By embracing the importance of effective data collection, you are embarking on a journey that empowers you to contribute meaningfully to research, decision-making, and the advancement of knowledge.

1.4 Understanding the Role of Tools and Instruments

Embarking on successful data collection endeavors hinges on understanding the pivotal role played by tools and instruments in this process. These tools serve as the vital conduits connecting researchers to the invaluable insights held within the data they seek to gather. They encompass an array of methodologies and mechanisms meticulously crafted to extract information systematically, ensuring the reliability and validity of collected data. Figure 1.1 provides a visual representation elucidating the functions of tools and instruments in data collection, offering insights into their significance and impact on research outcomes.

```
        Achieve Research
        **************|
          ***|        |
         ***  |        |
        ***    |        |
       ***      |        |
Minimize Bias |        | Standardize Data
       ***      |        |
        ***    |        |
         ***  |        |
          ***|        |
        **************|
          Capture Data
```

Figure 1.1. The Role of Tools and Instruments.
Source: Author

1.4.1 Defining Tools and Instruments

At the heart of effective data collection lie the essential constructs of tools and instruments, which serve as the conduits connecting researchers to the valuable insights held within the data. These entities encompass a diverse array of methodologies, mechanisms, and strategies meticulously crafted to systematically extract information from participants, subjects, or sources. They represent the bridge between raw data and meaningful knowledge, translating the complexities of human experiences, observations, and phenomena into quantifiable and comprehensible forms. Whether manifesting as tangible entities like questionnaires and sensors or taking conceptual form as interview guidelines and observation protocols, tools and instruments embody the art and science of data collection, transforming the abstract landscape of research into tangible insights that fuel discovery and understanding.

1.4.2 Functions of Tools and Instruments

The primary functions of tools and instruments in data collection are to:

Standardize Data Collection: Tools ensure that data is collected in a consistent and structured manner. This standardization enhances the reliability and validity of the collected information.

Capture Data: Instruments facilitate the capture of various types of data, including quantitative (numerical) and qualitative (descriptive) data. They provide a framework for systematically recording observations and responses.

Minimize Bias: Well-designed tools and instruments are formulated to minimize bias and subjectivity during data collection. This ensures that the information gathered is as objective and accurate as possible.

Facilitate Analysis: Effective tools and instruments generate data that can be readily analyzed. The structure and format of the collected data enable researchers to draw insights and conclusions.

Achieve Research Objectives: Different research questions require different types of data. Tools and instruments are tailored to specific research objectives, enabling researchers to gather the information they need.

1.4.3 Considerations for Tool and Instrument Selection

Selecting appropriate tools and instruments is a critical decision in the data collection process. Several factors influence this choice, including:

Research Objectives: The type of data needed and the research questions to be answered drive the selection of tools and instruments.

Population: Considerations about the characteristics of the population being studied, including demographics and context, influence the choice of tools.

Data Types: Different tools are suitable for collecting quantitative data (numbers) versus qualitative data (descriptions).

Methodology: The research methodology being employed, such as surveys, interviews, observations, or experiments, guides the selection of tools.

Resources: The availability of resources, including time, budget, and personnel, impacts the choice of tools and instruments.

1.4.4 The Iterative Process

The process of selecting and refining tools and instruments is often iterative. Researchers may need to adapt or modify their tools based on preliminary results, feedback from participants, or unexpected challenges encountered during data collection.

In the upcoming chapters of this guide, we will discuss the various data collection methods and explore the specific tools and instruments associated with each method. By understanding the role of tools and instruments, you'll gain insights into how these components contribute to the success of your data collection endeavors.

CHAPTER 2

Fundamentals of Data Collection

ABSTRACT
The second chapter lays the essential groundwork for effective research endeavors. It explores the distinctions between quantitative and qualitative data and elucidates the systematic steps integral to the data collection process. This foundational knowledge equips researchers with the insights needed to embark on successful data collection journeys.

Keywords: Data collection, research fundamentals, quantitative data, qualitative data, data collection process

2.1 Defining Goals of Data Collection

At the core of any research endeavor, the process of data collection plays a pivotal role in shaping the quality and validity of findings. Data collection is not merely the act of amassing information; it is a purposeful, systematic, and meticulous journey that aims to unearth insights, validate hypotheses, and contribute to the expansion of human knowledge. This section highlights the fundamentals of data collection, elucidating its multifaceted nature, significance, and the underlying goals that guide researchers in their quest for understanding.

2.1.1 The Goals of Data Collection

The objectives that steer data collection efforts are rooted in the pursuit of knowledge, accuracy, and meaningful interpretation. Several overarching goals underscore the significance of data collection in research:

2.1.1.1 Empirical Evidence and Informed Decision-Making

One primary goal of data collection is to provide empirical evidence that supports informed decision-making. By systematically collecting accurate and reliable data, researchers aim to present a factual foundation upon which stakeholders can base their choices. In scientific research, data serves as a reliable touchstone for validating hypotheses, offering tangible evidence that guides researchers toward conclusions grounded in reality.

2.1.1.2 Validating Research Hypotheses

Data collection serves as the litmus test for research hypotheses. Hypotheses, formulated based on existing knowledge and theories, require empirical validation to be considered credible. Through meticulous data collection, researchers can gather evidence that either confirms or refutes their hypotheses, leading to more accurate conclusions and insights.

2.1.1.3 Contributing to Knowledge Expansion

Data collection is the vessel through which researchers contribute to the expansion of human knowledge. By systematically gathering data, researchers can explore new frontiers, uncover hidden relationships, and add to the existing body of knowledge within their respective fields. Each data point serves as a building block that enriches the collective understanding of phenomena, fostering a continuous cycle of discovery and exploration.

2.1.1.4 Facilitating Comparative Analysis

Data collected using standardized methods allows for comparative analysis, enabling researchers to draw meaningful comparisons across different contexts, locations, or periods. This goal is instrumental in identifying trends, variations, and patterns that can inform decision-making, policy formulation, and future research directions.

2.1.1.5 Supporting Evidence-Based Practices

Data collection is the bedrock upon which evidence-based practices are built. In an era where evidence-driven decisions are paramount, reliable data serves as the cornerstone of policies, interventions, and strategies. By collecting and analyzing data rigorously, practitioners and decision-makers ensure that their actions are rooted in empirical evidence rather than conjecture.

In summary, data collection is a dynamic and purposeful endeavor that underpins research across disciplines. With its multifaceted goals of providing empirical evidence, validating hypotheses, expanding knowledge, facilitating comparisons, and supporting evidence-based practices, data collection forms the backbone of informed decision-making and the relentless pursuit of understanding. As the subsequent chapters of this book unfold, the intricacies of data collection tools, methods, and techniques will be explored, further illuminating the art and science of extracting valuable insights from the intricacies of our world.

2.2 Types of Data: Quantitative and Qualitative

Understanding the nature of data is a fundamental step in the data collection process. Data can be broadly classified into two main types: quantitative and qualitative. These two categories represent distinct approaches to gathering, analyzing, and interpreting information, each offering unique insights and perspectives that contribute to a comprehensive understanding of the research subject. This section delves into the characteristics, purposes, and applications of quantitative and qualitative data, highlighting their significance in research and decision-making. Table 2.1 provides a visual representation of the distinctions between quantitative and qualitative data, elucidating their characteristics, purposes, and applications in research and decision-making.

Table 2.1. Differences between Quantitative and Qualitative Data

Aspect	Quantitative Data	Qualitative Data
Definition	Numerical data that can be measured	Non-numerical data capturing nuances
Collection Methods	Surveys, experiments, measurements	Interviews, focus groups, observations
Precision	Numerical precision and accuracy	Descriptive detail and contextual richness
Objectivity	Objective, structured methods	Subjective, embraces subjectivity
Data Analysis	Statistical analysis	Thematic analysis
Contextual Understanding	Limited, focuses on numerical relationships	Rich, explores context and human experiences
Hypothesis Testing	Commonly used	Less common, theory generation
Applications	Financial analysis, market research, epidemiology	Sociology, psychology, anthropology

Source: Author

2.2.1 *Quantitative Data*

Quantitative data refers to information that is expressed in numerical form and can be quantified and measured. This type of data is characterized by its objectivity, as it deals with empirical observations and numerical values. Quantitative data is typically collected through structured methods such as surveys, experiments, and measurements. It is particularly well-suited for research scenarios where precise measurement, statistical analysis, and quantifiable relationships are essential.

2.2.1.1 Characteristics of Quantitative Data

Quantitative data exhibits a set of distinctive characteristics that distinguish it as a precise and objective form of data:

Numerical Precision: One of the fundamental attributes of quantitative data is its reliance on specific numerical values. These values are typically quantitative, such as counts, measurements, or percentages. This numerical precision offers several advantages. It allows researchers to perform precise measurements, make accurate comparisons, and perform mathematical operations on the data. For instance, when measuring the height of individuals in a population, the numerical values are not only specific but also provide a clear hierarchy for comparisons.

Objectivity: Quantitative data is inherently designed to be objective. This objectivity stems from the structured and standardized methods employed in its collection and analysis. Researchers strive to minimize personal bias or subjectivity when collecting quantitative data. By adhering to predefined protocols, using standardized instruments, and applying consistent measurement techniques, they aim to ensure that the data represents an impartial reflection of the phenomenon under study. For example, in a survey conducted to assess public opinion on a political issue, the questions are carefully crafted to avoid leading or biased wording, preserving the objectivity of the data.

Structured Methods: The collection of quantitative data often involves structured methods. Researchers use established frameworks, questionnaires, scales, or measurements to gather information systematically. These methods are designed to be uniform and consistent, allowing for the replication of data collection processes across different settings or periods. Structured data collection facilitates the comparability of data, as researchers can apply the same methods to different groups or populations. For instance, in a market research survey, a standardized questionnaire with closed-ended questions might be administered to gather quantitative data about consumer preferences. This structured approach ensures that respondents are presented with the same set of questions invariably.

Statistical Analysis: Quantitative data lends itself to rigorous statistical analysis. Statistical techniques enable researchers to derive deeper insights from the data by uncovering patterns, correlations, and trends that may not be immediately apparent. These analyzes can range from descriptive statistics that summarize data, such as mean and standard deviation, to inferential statistics that make predictions or test hypotheses. Statistical analysis adds a layer of objectivity and rigor to the interpretation of quantitative data. For instance, in a clinical trial studying the effectiveness of a new medication, statistical analysis can reveal whether the medication's impact on patients is statistically significant compared to a control group.

Discreteness and Continuity: Quantitative data can be either discrete or continuous. Discrete data consists of distinct, separate values, often counted in whole numbers (e.g., the number of cars in a parking lot). Continuous data, on the other hand, represents measurements that can take any value within a given range and can include decimals (e.g., temperature measurements in degrees Celsius). This distinction is important when selecting appropriate statistical techniques for analysis.

Replicability: Quantitative research often emphasizes the replicability of data collection processes. Researchers aim to design studies in a way that allows others to replicate their experiments or surveys using the same methods and instruments. This replicability enhances the reliability and credibility of the findings.

Precision and Accuracy: Quantitative data values are expected to be precise and accurate. Precision refers to the level of detail or granularity in the measurements, while accuracy refers to the degree of closeness between the measured value and the true value. Researchers take measures to minimize errors and ensure that the collected data accurately reflects the phenomenon being studied.

Scalability: Quantitative data is scalable, meaning that it can be collected from a small sample or a large population. This scalability allows researchers to conduct studies at various scales, from local surveys to nationwide censuses, and draw conclusions that apply to different levels of analysis.

Quantifiability: As the name suggests, quantitative data can be quantified, meaning it can be expressed in numerical terms. This quantifiability enables researchers to assign values to variables and measure relationships between them. It's particularly useful when dealing with variables that have clear numeric interpretations, such as age, income, or test scores.

Ease of Summarization: Quantitative data is amenable to summarization through various statistical and graphical methods. Researchers can create summary statistics like means, medians, and standard deviations to describe central tendencies and variability in the data. Visual representations such as histograms, box plots, and scatterplots provide insights into the distribution and patterns within the data.

Hypothesis Testing: Quantitative data is well-suited for hypothesis testing. Researchers can formulate hypotheses, design experiments or surveys to test these hypotheses and use statistical tests to determine whether the data supports or rejects the hypotheses. This hypothesis-driven approach is a hallmark of quantitative research.

Standardization: In many cases, quantitative data benefits from standardization, which ensures that measurements are consistent and comparable. Standardized scales and units of measurement are commonly used in quantitative research to facilitate data collection and analysis across different contexts.

2.2.1.2 Applications of Quantitative Data

Quantitative data, with its emphasis on precision and numerical representation, serves as a versatile tool across a wide range of fields and applications. Here are additional ways in which quantitative data is applied:

1. Quality Control and Assurance: In manufacturing and industry, quantitative data is crucial for quality control. Measurements of product dimensions, defects, and tolerances are used to ensure products meet specific standards and specifications.

2. Financial Analysis: Quantitative data is the backbone of financial analysis. It is used to evaluate the performance of financial assets, assess investment risks, and make informed decisions in stock markets, asset management, and investment portfolios.

3. Market Research: Quantitative data is employed to understand consumer behavior, preferences, and market trends. Surveys, questionnaires, and data analysis help businesses make strategic decisions in product development, pricing, and marketing.

4. Education Assessment: In educational settings, quantitative data plays a significant role in student assessments and performance evaluations. Standardized tests, grading systems, and educational statistics rely heavily on quantitative data to gauge student achievement and educational outcomes.

5. Environmental Monitoring: Environmental scientists collect quantitative data to monitor and assess environmental conditions. Parameters like air quality, water pollution levels, and biodiversity counts are quantified to inform conservation efforts and environmental policies.

6. Epidemiology and Public Health: Quantitative data is essential in tracking disease prevalence, analyzing health outcomes, and making public health recommendations. It aids in epidemiological studies, vaccine efficacy assessments, and health policy decisions.

7. Social Sciences: Social scientists use quantitative data to study human behavior and social phenomena. Surveys, experiments, and statistical analysis provide insights into topics like crime rates, voting behavior, and social inequality.

8. Engineering and Technology: Engineers rely on quantitative data for design, testing, and optimization. Quantitative data aids in structural analysis, product development, and troubleshooting in fields such as civil engineering, mechanical engineering, and computer science.

9. Sports Analytics: Quantitative data has revolutionized sports analysis, enabling teams and coaches to make data-driven decisions. Player performance metrics, game statistics, and play-by-play data are used for strategy development and player selection.
10. Political Science: Quantitative data informs political science research by providing data on election results, voter demographics, and policy impacts. It helps political scientists analyze the behavior of political actors and assess the effectiveness of policies.
11. Crime Analysis: Law enforcement agencies utilize quantitative data to study crime patterns, allocate resources effectively, and develop crime prevention strategies. Crime statistics, geographic information systems (GIS), and predictive policing rely on quantitative analysis.
12. Transportation Planning: Quantitative data assists transportation planners in optimizing traffic flow, designing efficient transportation systems, and assessing the impact of infrastructure projects. Traffic counts, travel time data, and congestion measurements contribute to informed decisions.

In essence, quantitative data is a cornerstone of empirical research and decision-making across numerous disciplines. Its versatility, precision, and ability to reveal patterns and relationships make it an invaluable tool for addressing complex challenges and gaining a deeper understanding of the world around us.

2.2.2 *Qualitative Data*

Qualitative data, in contrast, involves non-numerical information that captures the nuances, complexities, and context of a phenomenon. This type of data is collected through open-ended methods such as semi-structured or open-ended unstructured interviews, focus group discussions, observations, and content analysis. Qualitative data seeks to explore the depth and richness of human experiences, attitudes, perceptions, and behaviors, providing insights that are often difficult to capture using quantitative methods alone (Tenny et al., 2022; Mohajan, 2018).

2.2.2.1 Characteristics of Qualitative Data

Qualitative data is characterized by its unique qualities that make it particularly suited for exploring complex human experiences and phenomena. Here are further details on its defining characteristics:

1. Descriptive Detail: Qualitative data excels at capturing the richness and depth of human experiences. It goes beyond mere numerical measurements to provide descriptive, narrative accounts. Researchers using qualitative methods aim to understand the intricacies of a phenomenon by examining context, personal stories, and the social and cultural factors that influence behaviors and opinions. For example, in a qualitative study on the experience of refugees, researchers might collect detailed narratives to gain insight into their challenges, aspirations, and coping mechanisms.
2. Subjectivity: Unlike quantitative data, which strives for objectivity, qualitative data embraces subjectivity. It acknowledges that individuals bring their unique perspectives, values, and interpretations to the data collection process. Researchers often engage with the subjectivity of participants, recognizing that their viewpoints and emotions contribute to the narrative. This subjectivity is seen as a valuable source of insights, allowing researchers to explore diverse perspectives within a given context.
3. Open-Ended Methods: Qualitative data collection methods are typically open-ended and flexible. Rather than relying on structured questionnaires or predetermined measurements, qualitative researchers use methods like interviews, focus groups, participant observation, and open-ended surveys. These methods encourage participants to share their thoughts and experiences in their own words, providing a wealth of detailed information. Open-ended methods facilitate exploration and discovery,

allowing unexpected themes and insights to emerge during data collection.

4. Thematic Analysis: Qualitative data analysis involves identifying themes, patterns, and recurring concepts within the collected data. Researchers immerse themselves in the data, often through techniques like coding and categorization, to uncover underlying meanings and connections. Thematic analysis allows for the organization of qualitative data into coherent narratives or explanatory frameworks. Researchers identify themes that reflect the experiences, perspectives, and stories of participants, providing a deeper understanding of the phenomenon being studied.5. Contextual Understanding: Qualitative data excels at providing a contextual understanding of the phenomena under investigation. It considers the broader context within which behaviors, events, or experiences occur. This contextualization helps researchers appreciate the influence of culture, environment, history, and social dynamics on the data.

5. Holistic Approach: Qualitative research takes a holistic approach, examining the entire context rather than isolating specific variables. It explores how various elements interact and contribute to the complexity of the phenomenon. Researchers aim to understand the whole picture rather than focusing solely on discrete aspects.

6. Richness and Depth: Qualitative data is often described as rich and deep. It allows researchers to delve deeply into a topic, exploring nuances, contradictions, and subtle variations. This depth is particularly valuable when studying intricate subjects such as human emotions, perceptions, or social interactions.

7. Emergent Design: Qualitative research often follows an emergent design, where the research process is flexible and adaptive. Researchers may adjust their data collection strategies or research questions as new insights emerge. This flexibility allows

for the exploration of unexpected avenues and adaptability to the evolving research context.

8. Participant Engagement: Qualitative data collection methods often involve active engagement with participants. Researchers establish rapport, build trust, and encourage participants to share their stories and perspectives openly. This participant engagement can lead to richer data and a deeper understanding of the subject matter.

9. Diversity of Data Sources: Qualitative data can be derived from various sources, including interviews, focus groups, participant observation, documents, photographs, and audiovisual recordings. This diversity of data sources enables researchers to triangulate information, cross-validate findings, and gain a multifaceted view of the phenomenon.

10. Contextualization of Quantitative Data: Qualitative data can complement quantitative data by providing context and meaning to statistical findings. Qualitative insights help explain why certain quantitative patterns or correlations exist, shedding light on the "how" and "why" behind the numbers.

11. Reflexivity: Qualitative researchers often engage in reflexivity, acknowledging their role and potential biases in the research process. They reflect on their perspectives, assumptions, and preconceptions to maintain transparency and rigor. Reflexivity contributes to the researcher's self-awareness and the credibility of the study.

12. Narrative Approach: Qualitative data often takes a narrative approach, where participants' stories are central to the research. These narratives are not only valuable for understanding individual experiences but also for constructing a broader narrative that captures collective experiences or societal trends.

In essence, qualitative data is characterized by its depth, flexibility, contextual richness, and focus on understanding the complex interplay of human experiences and social phenomena. Researchers turn to qualitative methods when seeking to explore, interpret, and illuminate the multifaceted aspects of their research questions.

2.2.2.2 Applications of Qualitative Data

Qualitative data, with its capacity for in-depth exploration and contextual understanding, finds diverse and invaluable applications across various fields. Here's an elaboration on its applications:

Explore Phenomena: Qualitative data serves as a powerful tool for delving into complex phenomena across disciplines. Researchers employ qualitative methods to investigate intricate social, cultural, and behavioral patterns. For instance, in anthropology, qualitative research uncovers the intricacies of indigenous cultures and their adaptive strategies. In sociology, it unravels the dynamics of social movements and the factors that drive them. In psychology, qualitative data helps unravel the nuances of human emotions and behaviors. This exploration goes beyond surface-level observations, offering a profound comprehension of the multifaceted aspects that shape our world.

Capture Subjective Experiences: One of the distinctive strengths of qualitative data lies in its ability to capture the rich tapestry of human experiences. It provides a platform for individuals to express their feelings, perceptions, and perspectives authentically. Whether in the context of therapy sessions, where individuals share their emotional journeys or in the realm of literature and arts, where qualitative data sources like personal narratives and interviews inspire creative works that resonate with universal human experiences. By capturing these subjective dimensions, qualitative data enriches our understanding of the human condition.

Generate Theories: Qualitative data is not confined to describing existing phenomena; it also plays a pivotal role in theory generation. Researchers employ qualitative research to construct and refine theories across disciplines. For instance, in education, qualitative data may be used to develop innovative pedagogical theories by understanding how students engage with learning materials. In business and management, qualitative research informs theories of leadership by studying the behaviors and motivations of successful leaders. These emergent theories are often rooted in real-world observations and have the potential to reshape paradigms within their respective fields.

In essence, qualitative data's applications extend far beyond the boundaries of specific disciplines. It empowers researchers, practitioners, and creative minds to explore the depths of human existence, inspire change, and construct new theories that enhance our comprehension of the world around us.

In summary, the distinction between quantitative and qualitative data is pivotal in shaping the research process. Quantitative data offers numerical precision, objectivity, and statistical rigor, while qualitative data provides rich context, subjective insights, and the exploration of human experiences. The choice between these types of data depends on research goals, methodologies, and the depth of understanding sought, as both approaches contribute to a holistic and multifaceted comprehension of the research subject.

2.3 Steps in the Data Collection Process

The data collection process is a systematic and intricate journey that involves a series of well-defined steps to ensure the accuracy, reliability, and validity of the collected information. These steps provide a roadmap for researchers to navigate through the complexities of gathering data, from conceptualization to interpretation. This section outlines the essential stages of the data collection process, emphasizing the importance of meticulous planning, ethical considerations, and rigorous execution. Major steps are presented in Figure 2.1.

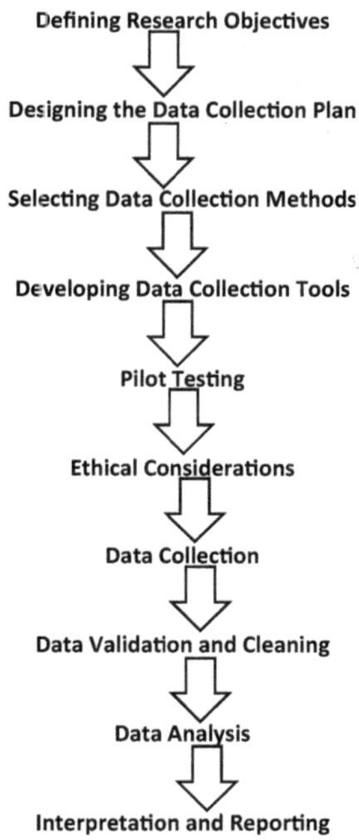

Figure 2.1. Steps in the Data Collection Process.

Source: Author

Step 1: Defining Research Objectives: The process commences with a clear articulation of research objectives and questions. This step involves identifying what information needs to be collected, the purpose of the data collection, and the expected outcomes. Research objectives act as guiding beacons that shape the entire data collection process.

Step 2: Designing the Data Collection Plan: A well-structured data collection plan is crucial to ensure the efficiency and effectiveness of the process. Researchers determine the appropriate data collection methods, tools, and instruments to be used. They consider factors such as the target population, data type (quantitative or qualitative), and logistics of data collection.

Step 3: Selecting Data Collection Methods: Based on research objectives, researchers choose the most suitable data collection methods. These methods can include surveys, interviews, observations, focus groups, experiments, and more. The choice of method aligns with the nature of the research, the type of data required, and the available resources.

Step 4: Developing Data Collection Tools: Data collection tools, such as questionnaires, interview guides, or observation protocols, are designed and structured to elicit the desired information. Tools must be clear, unbiased, and aligned with research objectives to ensure data accuracy and relevance.

Step 5: Pilot Testing: Before full-scale data collection, a pilot test is conducted to identify and rectify any issues with the data collection process. This involves testing the data collection tools, methods, and procedures on a small sample to assess their feasibility, clarity, and effectiveness.

Step 6: Ethical Considerations: Ethical considerations are paramount in data collection. Researchers must ensure the protection of participants' rights, privacy, and well-being. Informed consent is obtained from participants, and measures are taken to maintain confidentiality and anonymity, especially when dealing with sensitive topics.

Step 7: Data Collection: The actual data collection phase involves executing the planned methods and procedures. Researchers gather data from participants, subjects, or sources using the chosen tools. Attention to detail and adherence to the established protocols are critical to ensure the validity of the collected data.

Step 8: Data Validation and Cleaning: Upon data collection, researchers engage in data validation and cleaning. This involves checking for errors, inconsistencies, or outliers that might affect the quality of the data. Any discrepancies are addressed, and data is organized for analysis.

Step 9: Data Analysis: Data analysis transforms raw data into meaningful insights. Quantitative data may undergo statistical analyzes to uncover patterns and relationships, while qualitative data is subjected to thematic analysis to identify recurring themes and narratives.

Step 10: Interpretation and Reporting: In the final step, researchers interpret the analyzed data in the context of the research objectives. Conclusions are drawn, implications are discussed, and findings are reported clearly and concisely, often through research papers, presentations, or reports.

In summary, the data collection process is a meticulously planned and executed endeavor that spans multiple stages, from conceptualization to reporting. Each step is interdependent and contributes to the integrity and validity of the research outcomes. By adhering to ethical principles, employing appropriate data collection methods, and rigorously analyzing the gathered data, researchers ensure that their findings are accurate, relevant, and capable of contributing meaningfully to the body of knowledge in their respective fields.

CHAPTER 3

Selecting the Right Tools and Instruments

ABSTRACT
This chapter discusses the crucial decisions that researchers face in choosing the most appropriate data collection tools. It explores the considerations influencing tool selection, the art of aligning tools with research objectives, and an exploration of commonly employed data collection instruments. This chapter provides a strategic compass to navigate the diverse landscape of research methodologies.

Keywords: Tool selection, research instruments, data collection tools, research objectives, instrument choice

3.1 Considerations for Tool and Instrument Selection

The effectiveness and integrity of data collection heavily rely on the careful selection of appropriate tools and instruments. The process of choosing the right tools involves a nuanced evaluation of various factors to ensure that they align with research goals, capture accurate information, and yield meaningful insights. This section delves into the critical considerations researchers must take into account when selecting tools and instruments for data collection, highlighting the importance of alignment, validity, reliability, feasibility, and ethical considerations.

Alignment with Research Objectives: The first and foremost consideration when selecting tools and instruments is their alignment with the research objectives. Researchers must ask themselves how the chosen tools will contribute to addressing the research questions and objectives. A tool that accurately captures the desired data points and variables is essential for ensuring that the collected information is relevant and meaningful.

Validity and Reliability: Validity and reliability are key indicators of the quality of collected data. Validity ensures that the tool measures what it is intended to measure, while reliability ensures consistent results upon repeated measurements. Researchers should assess the validity and reliability of tools through pilot testing, pre-testing, and validation studies. This process helps identify any biases, ambiguities, or inconsistencies that might compromise the accuracy of the collected data.

Suitability for Data Type: Different research questions require different types of data—quantitative or qualitative. Researchers must ensure that the chosen tools are suitable for the type of data they aim to collect. For quantitative data, tools must allow for precise measurement and numerical analysis, while for qualitative data, tools should facilitate open-ended responses and nuanced exploration.

Feasibility and Resources: The practical feasibility of using a particular tool is a critical consideration. Researchers should assess whether the chosen tools can be effectively implemented within the available resources, such as time, budget, and personnel. Complex tools that require specialized training or extensive resources may not be feasible for certain research projects.

Ethical Considerations: Ethical considerations play a crucial role in tool selection. Researchers must ensure that the chosen tools respect participants' rights, privacy, and well-being. Tools should not cause discomfort, harm, or distress to participants. Ethical considerations also extend to the way data is collected, stored, and shared, emphasizing transparency and informed consent.

Cultural Sensitivity: Cultural sensitivity is paramount, especially in cross-cultural or international research. Researchers should evaluate whether the chosen tools are culturally appropriate and sensitive to the beliefs, values, and norms of the target population. Cultural mismatch between tools and participants can lead to biased or inaccurate data.

Expertise and Training: Researchers and data collectors need the necessary expertise to effectively use the chosen tools. Proper training ensures that data collection is carried out correctly, minimizing errors and biases. Researchers should consider the level of expertise required for tool implementation and whether they have access to training resources.

Pilot Testing and Iteration: Pilot testing is a crucial step in tool selection. Testing the tools on a small sample helps identify potential issues, refine the tools, and ensure that they are clear, unbiased, and effective. Iteration based on pilot testing feedback enhances the tools' quality and suitability.

Flexibility and Adaptability: Research environments can be dynamic, requiring tools that can adapt to changing conditions or unexpected circumstances. Tools that offer flexibility in their application and can accommodate modifications without compromising data quality are advantageous.

In summary, the process of selecting the right tools and instruments for data collection is a judicious endeavor that shapes the quality and integrity of research outcomes. By carefully considering alignment with research objectives, validity, reliability, data type, feasibility, ethics, cultural sensitivity, expertise, and adaptability, researchers ensure that their chosen tools are well-suited to capture accurate and meaningful data. This strategic selection process lays the foundation for robust and insightful data collection that contributes to the advancement of knowledge and informed decision-making.

3.2 Matching Tools to Research Objectives

Selecting the appropriate data collection tools involves a deliberate process of aligning tools with the specific research objectives. Each research question or hypothesis demands a tailored approach that maximizes the accuracy and relevance of the collected data. This section outlines the strategies and considerations researchers should keep in mind when matching tools to their research objectives, emphasizing the importance of precision, depth, and congruence between tools and the goals of the study.

Understanding Research Objectives: Before selecting tools, researchers must have a deep understanding of their research objectives. This involves clarifying the specific information they aim to gather, the variables they need to measure, and the type of insights they seek to derive from the data.

Quantitative Research Objectives: For research objectives that require quantifiable data, such as numerical measurements, statistical relationships, and patterns, quantitative data collection tools are essential. Surveys, questionnaires, experiments, and structured observations are commonly used tools in quantitative research. Researchers must ensure that the selected tools can precisely capture the variables of interest and facilitate accurate numerical analysis.

Qualitative Research Objectives: Qualitative research objectives, which focus on exploring complex phenomena, subjective experiences, and underlying motivations, necessitate qualitative data collection tools. Interviews, focus groups, participant observations, and content analysis are well-suited for capturing rich narratives, contextual insights, and diverse perspectives.

Mixed-Methods Research: In cases where research objectives demand a holistic approach that combines quantitative and qualitative data, mixed-methods research is employed. Researchers integrate both types of data collection tools to triangulate findings, validate results, and provide a comprehensive understanding of the research subject.

Depth and Complexity: The complexity of research objectives influences the depth and sophistication of data collection tools. Complex objectives may require tools that allow for open-ended responses, probing questions, and in-depth exploration. Simpler objectives may be addressed using structured tools with predefined response options.

Consideration of Variables: Researchers should identify the variables critical to their research objectives. This identification informs the design of tools that effectively capture these variables. For instance, if studying the impact of socio-economic factors on educational outcomes, researchers would design questions that gather information about income, parental education, and access to resources.

Data Sources and Participants: The nature of the data sources and participants also influences tool selection. Tools must be tailored to the characteristics of the population being studied, ensuring that questions are relevant, understandable, and culturally sensitive.

Alignment with Analytical Techniques: Researchers should consider the analytical techniques they intend to employ. Certain tools are better suited for specific analytical methods. For example, quantitative data collected through surveys can be subjected to statistical tests, while qualitative data from interviews can undergo thematic analysis.

Flexibility for Emerging Insights: While matching tools to research objectives, researchers should also allow room for unexpected insights. Flexibility in data collection tools enables researchers to adapt to discoveries, modify questions, and explore emerging themes.

In summary, matching tools to research objectives is a meticulous process that requires a deep understanding of the research goals and the nuances of data collection. By aligning tools with the nature of the research, variables of interest, complexity of objectives, and characteristics of participants, researchers ensure that the collected data is directly relevant, accurate, and capable of providing the insights necessary to address their research questions. This thoughtful alignment enhances the credibility and validity of the research outcomes, contributing to the broader understanding of the subject matter.

3.3 Exploring Common Data Collection Tools

In the realm of data collection, a diverse array of tools exists to cater to various research objectives, data types, and methodologies. This section delves into a selection of common data collection tools, shedding light on their characteristics, applications, and advantages within the context of research endeavors. From structured surveys to open-ended interviews, these tools offer researchers the means to gather valuable insights and contribute to the advancement of knowledge.

3.3.1 Surveys and Questionnaires

Surveys and questionnaires are structured data collection tools that employ a series of standardized questions to gather information from participants. These tools are versatile and can be administered through various mediums, such as paper-based forms, online platforms, or phone interviews. Surveys are particularly suited for quantitative research, as they allow for the efficient collection of numerical data and statistical analysis. They are used to gather demographic information, opinions, preferences, and attitudes.

3.3.2 *Interviews*

Interviews are a dynamic and personalized data collection tool that involves direct interaction between the researcher and participants. Interviews can be structured, semi-structured, or unstructured, depending on the research objectives. Structured interviews adhere to a predetermined set of questions, while semi-structured and unstructured interviews allow for open-ended exploration of topics. Interviews are valuable for gathering in-depth qualitative data, probing for nuances, and uncovering participants' perspectives and experiences.

3.3.3 *Observations*

Observations involve systematically watching and recording participants' behavior, actions, and interactions within specific contexts. This data collection tool is commonly used in naturalistic and field research, where researchers seek to understand social dynamics, cultural practices, and behavior in its natural setting. Observations can be participant-based, where the researcher engages in the observed activities, or non-participant, where the researcher remains an observer.

3.3.4 *Focus Groups*

Focus groups bring together a small group of participants to engage in guided discussions on a specific topic. This tool facilitates group interaction and encourages participants to share their viewpoints, opinions, and experiences. Focus groups are particularly useful for exploring complex social issues, capturing diverse perspectives, and generating insights through group dynamics and interactions.

3.3.5 *Content Analysis*

Content analysis involves systematic analysis of textual, visual, or audio content to identify patterns, themes, and meanings. This tool is used to analyze existing documents, media, or textual data to draw insights about attitudes, sentiments, or recurring topics. Content analysis is valuable for both quantitative and qualitative research, providing a structured framework to explore textual data.

3.3.6 *Experiments*

Experiments are controlled data collection methods used to investigate cause-and-effect relationships. Researchers manipulate independent variables to observe their impact on dependent variables while controlling for confounding factors. Experiments are common in natural and social sciences, allowing researchers to establish causal connections and test hypotheses rigorously.

3.3.7 *Diaries and Journals*

Diaries and journals are tools that enable participants to record their experiences, thoughts, and reflections over a specified period. These tools provide an unfiltered glimpse into participants' lives, emotions, and daily activities. Diaries and journals are often used in longitudinal studies or when researchers seek to understand the lived experiences of participants.

In summary, the selection of data collection tools depends on research objectives, data type, and the depth of exploration desired. Each tool offers unique advantages and insights, catering to the diverse landscape of research methodologies and objectives. By skillfully employing these tools, researchers gain access to a wealth of information that contributes to the understanding of complex phenomena, informs decision-making, and advances knowledge across various disciplines.

CHAPTER 4

Data Collection Tools: An Overview

ABSTRACT
This chapter provides an insightful exploration of diverse data collection methods commonly employed in research. It delves into the nuances of interviews, observations, focus group discussions (FGDs), key informant interviews (KIIs), community mapping, participatory rural appraisal (PRA), discourse analysis, and data collection from secondary sources. This chapter equips researchers with a comprehensive understanding of these methods, enabling informed choices in their research endeavors.

Keywords: Data collection methods, research techniques, interviews, observations, focus group discussions, key informant interviews, community mapping, participatory rural appraisal, discourse analysis, secondary data sources

4.1 Interviews

Data collection methods are the cornerstones of research, providing avenues to explore, understand, and analyze the world around us. Among these methods, interviews stand as a dynamic and personalized approach to gathering insights directly from participants. This section delves into the intricacies of interviews as a data collection method, encompassing their definition, various types, the process of conducting interviews, their popularity in research, as well as their merits and demerits.

4.1.1 Definition of Interviews

An interview is a purposeful and structured conversation between a researcher and one or more participants, with the aim of eliciting information, perspectives, and experiences related to the research objectives. Mason (2017) defines a research interview as an "in-depth, semi-structured, or loosely structured form of interaction with a specific purpose." To put it simply, a research interview consists of an interviewer (typically the researcher) asking a series of questions (Hussein, 2022). Interviews allow researchers to delve into participants' thoughts, emotions, opinions, and experiences in their own words, providing valuable qualitative data that can illuminate complex phenomena.

4.1.2 Types of Interviews

Interviews come in various types, each tailored to specific research goals and participant characteristics:

4.1.2.1 Structured Interviews

Structured interviews are a highly systematic method of data collection, where the interviewer asks a fixed set of predetermined questions in a specific order. This format ensures that each participant receives the same questions, posed in the same way, minimizing interviewer bias and allowing for consistent data collection across participants. The rigidity of structured interviews makes them particularly valuable in quantitative research and large-scale surveys, where the goal is to gather comparable data from a broad audience. By maintaining a strict adherence to the interview script, structured interviews provide reliable and objective data, facilitating statistical analysis and generalization of findings. They are commonly used in job recruitment processes, standardized surveys, and any context where uniformity and comparability of responses are crucial. Despite their lack of flexibility, which limits the exploration of unexpected topics, the precision and consistency of structured interviews make them an essential tool for gathering accurate, quantifiable data.

4.1.2.2 Semi-structured Interviews

Semi-structured interviews blend the rigidity of structured interviews with the flexibility of unstructured formats, offering a balanced approach to data collection. This type of interview follows a guided format, where a set of open-ended questions or topics are predetermined, yet the interviewer retains the flexibility to explore topics that emerge during the conversation. This adaptability allows the interviewer to delve deeper into unexpected areas of interest, providing richer, more nuanced data that captures the complexities of the subject matter.

Semi-structured interviews are particularly valuable in qualitative research, where the goal is to understand participants' perspectives and experiences in depth. They are commonly used in fields such as social sciences, health research, and market research, where exploring subjective viewpoints and obtaining detailed narratives are essential. The interviewer's ability to probe further based on responses leads to a more comprehensive understanding of the participants' thoughts and feelings, making semi-structured interviews a powerful tool for uncovering insights that structured interviews might miss.

By combining structure with flexibility, semi-structured interviews offer the best of both worlds: the consistency needed for comparing responses across participants and the openness required to capture the richness of individual experiences. This format is ideal for research scenarios where a balance between standardized data collection and the exploration of new insights is necessary.

4.1.2.3 Unstructured Interviews

Unstructured interviews are a highly flexible and open-ended form of interviewing where there is no predetermined set of questions or format. Instead, the interview is more of a conversation, allowing participants to discuss topics that are important to them in their own way and at their own pace. This format places the interviewee at the center, encouraging them to lead the dialog and express their thoughts and experiences freely.

The interviewer takes on a facilitative role, guiding the conversation only when necessary to explore certain themes or topics in more depth.

The lack of structure in these interviews allows for the exploration of complex and nuanced issues that structured approaches might overlook. Unstructured interviews are particularly valuable in qualitative research where the goal is to gain deep insights into individual experiences, opinions, and behaviors. They are commonly used in fields such as anthropology, psychology, and sociology, where understanding the context and meaning behind participants' words is crucial.

This type of interview is effective for exploring new areas of research, generating hypotheses, or gaining a comprehensive understanding of a subject from the participant's perspective. However, the open-ended nature of unstructured interviews can also pose challenges, such as difficulty in comparing responses across participants and the potential for interviewer bias. Despite these challenges, unstructured interviews remain a powerful tool for capturing rich, detailed data that reflects the complexity of human experiences.

4.1.2.4 Focus Group Interviews

Focus group interviews are a qualitative research technique that involves gathering a small group of people to discuss a specific topic or set of issues. Unlike one-on-one interviews, focus groups leverage the dynamic and interactive nature of group discussions to generate rich data and insights. These discussions are typically guided by a moderator who facilitates the conversation, encourages participation from all members, and keeps the discussion focused on the research objectives.

In a focus group, the interplay of ideas and opinions among participants can reveal a wide range of perspectives and stimulate new thoughts and concepts that might not emerge in individual interviews. The group setting can also provide a supportive environment where participants feel more comfortable expressing their views, especially when they share similar backgrounds or experiences. This method is particularly useful for exploring how individuals think and feel about a topic, as well as understanding the diversity of opinions within a community or target population.

Focus group interviews are commonly used in market research, social sciences, and policy research to gather information on consumer preferences, social attitudes, and community needs. They are also valuable for pretesting concepts, products, or services by eliciting feedback that can inform development and improvement. The insights gained from focus groups can be instrumental in shaping strategies, creating more effective communication, and understanding complex issues from multiple perspectives.

Despite their many advantages, focus group interviews also come with challenges. The data can be complex to analyze due to the conversational nature of the discussions, and there is a risk of dominant participants influencing the group's responses. Additionally, the moderator's skills in managing the group and guiding the discussion are crucial for obtaining high-quality data.

4.1.2.5 Behavioral Interviews

Behavioral interviews are a structured interview technique that focuses on understanding a candidate's past behavior as a predictor of future performance. Unlike traditional interviews that may ask hypothetical questions, behavioral interviews delve into specific instances where the candidate demonstrated skills or competencies relevant to the job. The underlying premise is that past behavior is the best indicator of how someone will perform in similar future situations.

In a behavioral interview, the interviewer typically asks the candidate to describe past experiences in detail, focusing on situations where they had to demonstrate certain skills, overcome challenges, or achieve significant outcomes. Questions often begin with phrases like, "Tell me about a time when you..." or "Give me an example of...". This approach helps to assess not only the candidate's competencies but also their thought processes, decision-making skills, and ability to handle various situations.

Behavioral interviews are particularly valuable in roles that require problem-solving, leadership, teamwork, and the ability to manage change or conflict. They are designed to elicit detailed responses that reveal how candidates have handled specific work situations in the past, providing insights into their suitability for the role. By focusing on concrete examples, behavioral interviews help to reduce the subjectivity and potential biases that can occur in more traditional interview formats.

Interviewers often use the STAR method (Situation, Task, Action, Result) to frame their questions and evaluate responses. This method helps candidates structure their answers in a way that clearly describes the context, their responsibilities, the actions they took, and the outcomes of their efforts.

4.1.2.6 Narrative Interviews

Narrative interviews are qualitative research methods designed to elicit rich, detailed accounts of personal experiences, stories, or life histories from participants. Unlike structured or semi-structured interviews that follow a predefined set of questions, narrative interviews are flexible and open-ended, allowing participants to narrate their stories in their own words. The focus is on capturing the complexities of individual experiences, perspectives, and interpretations of events. Researchers often use probing questions to encourage participants to delve deeper into their narratives, exploring themes, emotions, and meanings embedded within their stories. Narrative interviews are valuable for uncovering personal insights, cultural perspectives, and contextual understandings that may not be accessible through other research methods. They are widely used in fields such as psychology, sociology, anthropology, and education to explore identity, memory, social interactions, and the construction of personal and collective narratives.

4.1.2.7 Case Study Interviews

Case study interviews involve in-depth investigations of a particular individual, group, organization, or phenomenon within its real-life context. These interviews are structured to gather detailed information about specific cases, aiming to understand complex issues, behaviors, processes, or outcomes. Researchers use case study interviews to explore unique situations, uncovering rich data through multiple sources such as interviews, observations, documents, and archival records. The focus is on holistic examination, capturing the intricacies and dynamics of the case under study. Case study interviews are valuable in fields like business, education, medicine, and social sciences, providing insights into decision-making, problem-solving, organizational dynamics, and individual or group behaviors within their natural environments. They enable researchers to explore context-specific factors, interactions, and outcomes, contributing to both theoretical insights and practical applications.

Case study interviews are essential in research methodologies for their focused exploration of specific cases within real-world contexts. They involve comprehensive investigations into individuals, groups, organizations, or phenomena, aiming to uncover detailed insights and understand complex dynamics. Researchers employ various data sources such as interviews, observations, and documents to gather in-depth information. These interviews are particularly valuable in disciplines like business, education, medicine, and social sciences, where they shed light on decision-making processes, organizational behaviors, case histories, and the impact of interventions. By delving deeply into unique cases, case study interviews provide nuanced understandings that contribute both theoretical advancements and practical implications for diverse fields of study.

4.1.2.8 Ethnographic Interviews

Ethnographic interviews are integral to ethnographic research, involving immersive and in-depth interactions with individuals or groups within their cultural or social contexts. These interviews are characterized by their flexible and open-ended nature, allowing researchers to explore participants' perspectives, beliefs, behaviors, and social dynamics. Ethnographers often employ a participant-observation approach alongside interviews to gain a holistic understanding of the culture or community under study. These interviews aim to uncover underlying cultural norms, values, and practices that shape participants' experiences and interactions. Ethnographic interviews are widely used in anthropology, sociology, and cultural studies, providing rich, qualitative data that illuminates cultural phenomena and social processes from an insider's perspective.

Ethnographic interviews play a pivotal role in capturing the lived experiences and cultural nuances of individuals within their natural settings. Unlike more structured interview formats, ethnographic interviews are characterized by their fluidity and adaptability to the unfolding context of the fieldwork. Researchers engage deeply with participants, often over extended periods, fostering trust and rapport to elicit rich, detailed narratives. These interviews are designed to uncover tacit knowledge, local meanings, and social relationships that may not be apparent through other research methods. Ethnographers embrace the complexity and variability inherent in human behavior, using interviews to document how individuals navigate and interpret their social worlds. This approach allows for a nuanced exploration of cultural diversity, identity formation, and the intricate webs of meaning that shape community life. Ethnographic interviews thus serve as a powerful tool for generating thick descriptions and theoretical insights that contribute to our understanding of human societies and cultural dynamics.

4.1.2.9 Panel Interview

Panel interviews are structured sessions where a candidate or interviewee interacts with multiple interviewers simultaneously. This format allows organizations to gather diverse perspectives on a candidate's suitability for a role while providing a comprehensive evaluation of their skills, qualifications, and fit within the organizational culture. Typically, panel interviews involve a panel of two or more interviewers who may represent different departments, levels of seniority, or functional areas within the organization. Each panel member brings unique expertise and viewpoints to the interview process, enriching the assessment by exploring different facets of the candidate's background and experiences.

Panel interviews are known for their efficiency in assessing candidates across various dimensions. They enable interviewers to probe deeper into specific aspects of a candidate's resume, skills, and behavioral competencies through a structured questioning approach. Moreover, panel interviews offer a holistic view of how candidates respond to different personalities and perspectives, providing insights into their communication skills, adaptability, and ability to engage with diverse stakeholders. This format also allows interviewers to observe how candidates manage pressure, articulate their ideas, and demonstrate leadership qualities in a group setting. Overall, panel interviews are valued for their rigor and ability to yield a comprehensive evaluation that informs hiring decisions effectively.

4.1.2.10 In-depth Case Interviews

In-depth case interviews are a rigorous method used primarily in management consulting and similar fields to evaluate candidates' problem-solving abilities, analytical skills, and decision-making processes. Unlike traditional interviews that focus on assessing qualifications and experience, case interviews present candidates with real or hypothetical business scenarios, often drawn from the consulting firm's past projects or industry challenges. Candidates are expected to analyze the given information, identify key issues, develop logical frameworks, and propose actionable solutions within a limited time frame.

These interviews are structured to simulate the dynamic and fast-paced nature of consulting work, where problem-solving under pressure and effective communication are crucial. Interviewers typically assess how candidates structure their approach, think critically, and communicate their ideas clearly and persuasively. Success in case interviews hinges on a combination of analytical prowess, business acumen, creativity, and the ability to apply structured problem-solving frameworks such as SWOT analysis, Porter's Five Forces, or market sizing techniques. Candidates who excel in in-depth case interviews demonstrate not only their technical expertise but also their capacity to navigate complex challenges and provide strategic insights that add value to client organizations.

In summary, Table 4.1 provides a comprehensive overview of various interview types, their distinctive characteristics/features, and their appropriate contexts for applications/use.

Table 4.1. Characteristics and Applications of Different Types of Interviews

Type of Interview	Characteristics	Applications
Structured Interviews	Fixed set of questions; Same order for all participants; High consistency and reliability	Quantitative research; Surveys and large-scale studies; Job interviews and screening processes
Semi-Structured Interviews	Predefined questions; Allows for probing; Flexible and adaptable	Qualitative research; Case studies; Exploratory research
Unstructured Interviews	No predetermined questions; Conversational style; Deep exploration	Ethnographic studies; In-depth case studies; Exploratory interviews
Focus Group Interviews	Group discussion; Facilitator guides conversation; Interaction among participants	Market research; Product feedback; Social science research
Behavioral Interviews	Focus on past behaviors; Specific situational questions; Predict future behavior	Job interviews; Competency assessments; Psychological studies
Narrative Interviews	Storytelling format; Rich, contextual data; Focus on experiences	Oral histories; Life story research; Therapy and counseling sessions
Case Study Interviews	In-depth exploration; Focus on specific case; Detailed data collection	Case study research; Business and organizational analysis; Policy analysis
Ethnographic Interviews	Conducted in natural setting; Observational; Contextual understanding	Cultural research; Social anthropology; Community studies
Panel Interviews	Multiple interviewers; Diverse perspectives; Structured or semi-structured	Academic and job selection panels; High-stakes interviews; Policy and decision-making research
Depth Interviews	Intensive one-on-one; Detailed exploration; In-depth understanding	Psychological research; Market research; Complex issue exploration

Source: Author

4.1.3 When Do We Use Interview Tools for Data Collection?

Interviews are used as a data collection tool in various research contexts and scenarios where researchers aim to gather information directly from individuals or groups. Hussein (2022) highlights that interview methods are effective for gathering data on a wide range of topics. Some of these topics involve sensitive issues like age, education, race, marital status, social class, and income, where respondents might feel uneasy and may tend to overstate or understate their responses. Other issues, while not necessarily sensitive or threatening, could still be significant to respondents, leading them to provide detailed and thoughtful answers. On the other hand, some issues might be of lesser interest to the respondent compared to the researcher, and in such cases, the researcher may need to invest additional effort to obtain meaningful responses. Here are common situations when interviews are employed:

Quantitative Research: Interviews are used for quantitative data collection when researchers seek numerical responses to structured questions. This approach involves asking participants predetermined, closed-ended questions with fixed response options. It is commonly employed in surveys, questionnaires, assessments, and structured research interviews to gather standardized quantitative data.

Qualitative Research: Interviews are a fundamental tool in qualitative research, where researchers seek to understand complex human experiences, behaviors, and perspectives. Qualitative interviews are used in fields such as sociology, psychology, anthropology, and education to explore themes, narratives, and subjective viewpoints.

Exploratory Research: Interviews are often conducted in the early stages of research when there is limited existing knowledge about a topic. Researchers use exploratory interviews to generate hypotheses, identify key variables, and refine research questions.

In-Depth Understanding: Interviews allow researchers to gain in-depth insights into participants' thoughts, feelings, and motivations.

They are used to explore underlying meanings, beliefs, and the context of individual experiences.

Case Studies: In case study research, interviews are valuable for collecting rich, detailed information about specific cases or individuals. Researchers use interviews to construct comprehensive and contextually rich case narratives (for details, please see Yin, 2009).

Policy and Program Evaluation: Interviews are employed to assess the impact of policies, programs, or interventions on individuals or communities. Researchers gather feedback, opinions, and experiences to inform decision-making.

Clinical and Healthcare Research: Interviews are used to collect patient histories, assess symptoms, and understand the impact of medical conditions on individuals' lives. They are also employed in clinical trials and medical research.

Market Research: In the business world, interviews are used to gather customer feedback, preferences, and opinions. This information informs product development, marketing strategies, and customer satisfaction efforts.

User Experience (UX) Research: UX designers and researchers use interviews to understand users' needs, preferences, and pain points. This information guides the design of user-friendly products and interfaces.

Social and Political Research: Interviews are conducted in social and political research to gather data on public opinions, voting behavior, and attitudes toward various issues. They are also used in election polling.

Employment and Human Resources: Interviews are a standard tool for hiring and recruitment, enabling employers to assess candidates' qualifications, skills, and cultural fit.

Crisis Intervention: Mental health professionals use interviews to assess individuals in crisis and provide appropriate interventions or referrals.

Legal Proceedings: Attorneys and legal professionals conduct interviews to gather evidence, assess witnesses, and prepare for trials or legal cases.

Historical Research: Interviews with eyewitnesses or individuals with relevant knowledge are essential in historical research to gather firsthand accounts and perspectives on historical events.

Longitudinal Studies: Interviews are conducted at multiple time points to track changes and developments in individuals' lives over an extended period.

In these and other research and professional contexts, interviews offer a flexible and adaptable data collection method that allows researchers to engage with participants, ask open-ended questions, and explore complex topics. Interviews can be conducted in various formats, including structured, semi-structured, and unstructured, depending on the research goals and design.

4.1.4 How to Conduct Interviews

The process of conducting interviews is a multifaceted journey that involves careful planning, ethical considerations, effective communication, and meticulous data handling. Here's an expanded breakdown of the stages involved:

1. Preparation:
   Research Objectives: Before conducting interviews, researchers meticulously define their research objectives. They clarify what they aim to explore, understand, or uncover through the interviews. These objectives serve as the guiding light throughout the process.
   Interview Guides: Researchers develop interview guides or protocols tailored to their research objectives. These guides outline the key questions, topics, or themes that will be covered during the

interviews. They also include probing questions to elicit in-depth responses.
Participant Recruitment Strategies: Researchers devise strategies to recruit suitable participants who meet the criteria relevant to the research objectives. Whether through random sampling, purposive sampling, or snowball sampling, the recruitment process is aligned with the research goals.

2. Recruitment:
   Participant Selection: Participants are carefully selected based on the defined criteria. This selection ensures that the interviewees possess the knowledge, experiences, or perspectives necessary to contribute meaningfully to the research.
   Inclusivity: Researchers consider inclusivity, striving to include diverse voices and perspectives to enrich the data and minimize bias. This may involve reaching out to individuals from various backgrounds or demographics.
3. Informed Consent:
   Ethical Considerations: Ethical principles underpin the interview process. Researchers prioritize informed consent, ensuring that participants fully understand the research's purpose, potential risks, benefits, and how their data will be used. They address any questions or concerns participants may have.
   Confidentiality: Researchers assure participants that their identities and responses will be kept confidential, promoting a safe environment for open sharing.
4. Conducting the Interview:
   Interview Type: Researchers conduct the interviews according to the chosen interview type, which may include structured interviews with predefined questions, semi-structured interviews allowing for flexibility and follow-up questions, or unstructured interviews providing open-ended conversations.
   Active Listening: During the interview, active listening is paramount. Researchers attentively listen to participants, fostering a sense of trust and rapport. They validate participants' perspectives through verbal and non-verbal cues.

Adherence to Guide: Researchers follow the interview guide while remaining adaptable to participants' responses. They use probing techniques to dig deeper into topics and encourage participants to share their insights.

5. Recording:
   Data Capture: Interviews are recorded to ensure the accurate capture of data. Recording methods may include note-taking, audio recordings, or video recordings, depending on what is most suitable for the research context.
   Documentation: Researchers maintain detailed documentation, including date, time, location, and any contextual observations that may enhance data analysis.

6. Transcription:
   Transcription Process: Recorded interviews are transcribed into text format, converting spoken words into written documents. Transcription ensures that the data is ready for analysis. Researchers or trained professionals may handle this task.

The process of conducting interviews is a dynamic and interactive journey where researchers navigate ethical considerations, build rapport, and extract valuable insights from participants. It is through this process that the richness of qualitative data unfolds, contributing to the advancement of knowledge in various fields.

4.1.5 Popularity of Interviews in Research

Indeed, interviews have garnered popularity in research for several compelling reasons, making them a favored data collection method:

1. Rich and Contextual Insights: Interviews offer a platform for participants to express themselves in their own words. This allows researchers to access rich, context-specific information that may not be captured through quantitative methods alone. Participants can provide detailed narratives, anecdotes, and personal experiences, which add depth and meaning to the data.

2. Probing and Clarification: Unlike surveys or questionnaires, interviews enable researchers to probe and seek clarification. Researchers can ask follow-up questions, request examples, and delve deeper into participants' responses. This dynamic interaction helps uncover nuances and provides a comprehensive understanding of the topic.
3. Flexibility: Interviews come in various forms, such as structured, semi-structured, and unstructured. This flexibility allows researchers to adapt the interview style to the research objectives and the nature of the participants. It accommodates both quantitative and qualitative research needs.
4. Participant Viewpoints: Interviews prioritize the voices and perspectives of participants. Researchers can gain direct access to participants' viewpoints, beliefs, and emotions. This participant-centered approach helps in uncovering hidden or less-explored aspects of a subject.
5. Complex Topics: For research topics that are multifaceted, sensitive, or complex, interviews are particularly valuable. They allow researchers to navigate intricate terrain, such as personal values, cultural influences, and social dynamics, with a level of depth and sensitivity that quantitative methods may struggle to achieve.
6. Real-Time Insights: Interviews offer real-time insights, especially when conducted face-to-face or via video conferencing. Researchers can observe participants' body language, facial expressions, and non-verbal cues, which can provide additional layers of understanding.
7. Empirical Validation: Interviews help validate or refute research hypotheses. Through in-depth conversations, researchers can gather empirical evidence to support or challenge their initial assumptions, contributing to the rigor of the research.
8. Tailored Approach: Interviews can be tailored to specific participants or groups. This customization ensures that the questions and prompts are relevant and resonate with the individuals being interviewed, enhancing the quality of the data.

9. Human Connection: The interview process fosters a human connection between the researcher and the participant. This rapport can lead to more honest responses, as participants feel heard and valued.
10. Qualitative Depth: Interviews contribute to qualitative depth in research. They generate data that can be analyzed thematically, allowing researchers to identify patterns, themes, and emergent concepts, which can further inform theory development.

In sum, interviews stand as a cornerstone of qualitative research, valued for their ability to unearth the intricate layers of human experiences and provide researchers with a nuanced understanding of complex topics. Their adaptability and participant-centered approach make them a versatile and popular choice in the research landscape.

4.1.6 Merits of Interviews

The use of interviews as a data collection tool is possibly the most popular and widely employed method among researchers. Interviews as a data collection method come with a multitude of merits that make them a valuable tool in research. Here are some of the key advantages:

1. Rich Data: Interviews provide access to rich and in-depth data. Participants can express themselves freely, offering detailed insights, personal experiences, and context-specific information.
2. Contextual Understanding: Researchers can gain a deep understanding of the context in which participants operate. This contextual insight is particularly valuable for exploring complex topics or phenomena.
3. Flexibility: Interviews come in various formats, including structured, semi-structured, and unstructured. This flexibility allows researchers to adapt the interview style to the research objectives and participant characteristics.

4. Probing and Clarification: Researchers can ask follow-up questions and seek clarification during interviews. This probing ability helps uncover nuances, contradictions, or areas that require further exploration.
5. Participant-Centered: Interviews prioritize the perspectives and voices of participants. This approach allows researchers to respect and acknowledge the unique viewpoints of individuals.
6. Personal Connection: The interview process fosters a personal connection between the researcher and the participant. This rapport can lead to more open and candid responses.
7. Empirical Validation: Interviews allow researchers to gather empirical evidence to validate or challenge research hypotheses. They contribute to the rigor of the research by providing real-world data.
8. Real-Time Insights: In face-to-face or video interviews, researchers can observe participants' non-verbal cues, such as body language and facial expressions, adding depth to the data.
9. Tailored Questions: Researchers can customize questions and prompts to suit the specific needs and characteristics of the participants. This tailored approach ensures relevance and resonance.
10. Qualitative Depth: Interviews generate qualitative data that can be analyzed thematically. Researchers can identify patterns, themes, and emergent concepts, contributing to theory development.
11. Sensitive Topics: For sensitive or emotionally charged topics, interviews provide a safe and supportive environment for participants to share their experiences.
12. Real-Life Examples: Participants can provide real-life examples and anecdotes, making the data relatable and memorable.
13. Multiple Perspectives: Interviews allow researchers to collect data from multiple perspectives, offering a holistic view of the topic.

14. Hypothesis Testing: Through interviews, researchers can test hypotheses and gather evidence to support or refute their initial assumptions.
15. Researcher's Presence: The researcher's presence can enhance data quality by facilitating rapport, ensuring clarity, and adapting to unexpected insights.

In summary, interviews offer a dynamic and participant-centered approach to data collection, enabling researchers to access rich and nuanced information that may be challenging to obtain through other methods. Their adaptability, depth, and ability to explore complex topics make them a valuable asset in the research toolkit.

4.1.7 Demerits of Interviews

While interviews offer numerous advantages in data collection, they also come with some inherent limitations and demerits:

Subjectivity: Interviews are susceptible to researcher bias and subjectivity. The way questions are framed, the tone of voice, and the interviewer's body language can influence participants' responses.

Response Bias: Participants may provide socially desirable answers, especially in face-to-face interviews, leading to response bias. They might modify their responses to align with societal norms or to create a favorable impression.

Interviewer Bias: Interviewers may unintentionally influence participants' responses through their own biases or preconceptions. This can affect data quality and objectivity.

Resource-Intensive: Interviews can be resource-intensive in terms of time, effort, and cost. They require careful planning, conducting, and transcribing, which can be burdensome for researchers.

Small Sample Size: Compared to surveys, interviews typically involve a smaller sample size due to resource constraints. This limits the generalizability of findings to a broader population.

Interviewer Variability: In studies involving multiple interviewers, there may be variability in how interviews are conducted, potentially leading to inconsistencies in data collection.

Response Fatigue: Lengthy interviews or a high number of open-ended questions can lead to response fatigue, causing participants to provide less thoughtful or detailed responses as the interview progresses.

Limited Quantification: Interviews generate qualitative data, which can be challenging to quantify and analyze statistically. This limits the ability to draw quantitative conclusions.

Time-Consuming Analysis: Transcribing and analyzing interview data can be time-consuming, especially in studies with numerous interviews. Thematic analysis and coding require careful attention to detail.

Ethical Concerns: Ethical considerations are crucial in interviews. Maintaining confidentiality, obtaining informed consent, and ensuring participant well-being can be challenging but essential aspects of the process.

Difficulty with Sensitive Topics: Delving into sensitive or personal topics can be emotionally taxing for both participants and interviewers. Researchers must handle such situations with care and sensitivity.

Difficulty with Hard-to-Reach Populations: Interviewing certain populations, such as marginalized or vulnerable groups, may be challenging due to distrust, accessibility issues, or language barriers.

Cultural Bias: Cross-cultural interviews can be affected by cultural biases, misinterpretations, or misunderstandings that may impact the validity of responses.

Lack of Anonymity: In face-to-face interviews, participants may be hesitant to share sensitive information due to the lack of anonymity, which is more common in self-administered surveys.

Risk of Social Desirability Bias: Participants may provide responses they believe the interviewer expects, rather than their genuine opinions or experiences.

Despite these demerits, interviews remain a valuable qualitative research method, especially when exploring complex, nuanced topics or seeking in-depth insights. Researchers must be aware of these limitations and take measures to minimize their impact on data quality and reliability.

In summary, interviews serve as a versatile data collection method that offers researchers direct access to participants' thoughts, experiences, and perspectives. Through various interview types, researchers can adapt their approach to align with research objectives, generating qualitative data that contributes to a holistic understanding of complex phenomena. However, researchers must navigate challenges such as subjectivity and resource constraints while reaping the benefits of profound insights interviews offer.

4.2 Observations

Observation is a data collection method rooted in direct observation and recording of participants' behaviors, interactions, and occurrences within specific contexts. Hence, numerous discoveries and inventions in the modern world stem from the observations made by scientists (Farid, 2022). This method provides researchers with an unfiltered lens into real-life situations, enabling them to comprehend social dynamics, human behavior, and contextual influences. This section delves into the nuances of observations as a data collection method, encompassing their definition, types, the process of conducting observations, their popularity in research, as well as their merits and demerits.

4.2.1 Definition of Observations

Observations involve systematically watching and recording participants' behavior, actions, and interactions within particular settings. This method strives to capture spontaneous and authentic behaviors, devoid of researcher bias or influence, offering valuable insights into how individuals navigate their environment. Payne and Payne (2004) argue that pure observation, which involves solely watching people, is seldom employed in social research. This is primarily because human behavior is too intricate to be comprehensively documented through mere observation. Furthermore, strict observation tends to distance researchers from the subjects under study, limiting their ability to actively participate in or engage in meaningful conversations and interviews to gain deeper insights and understandings.

4.2.2 Types of Observations

Participant Observation: In this type, the researcher actively immerses themselves in the observed activities, essentially becoming a part of the environment being studied. This approach provides an insider's perspective and fosters rapport with the participants, often leading to deeper insights due to the researcher's direct involvement.

Non-Participant Observation: In contrast to participant observation, non-participant observation involves the researcher remaining a passive observer. They do not interact directly with the participants. This approach can yield more objective observations but may lack the depth of understanding offered by participant observation.

Structured Observation: Structured observations entail researchers following a predetermined checklist or coding scheme to systematically record specific behaviors or events. This approach is particularly useful when the research objective requires standardized data collection, making it easier to compare observations across different contexts or periods.

Unstructured Observation: Unstructured observations involve open-ended exploration of the research environment. Researchers do not adhere to a fixed checklist but instead have the freedom to record a wide range of behaviors and interactions as they occur. This approach offers flexibility and adaptability, allowing unexpected insights to emerge during the observation process.

Controlled and non-controlled observation: Controlled observation, as a method in social research, is characterized by its use of pre-arranged precision instruments, standardized conditions, and meticulous control over the phenomena or behaviors under examination. Typically conducted in laboratory settings, especially in experimental research, controlled observation prioritizes scientific accuracy (Lavrakas, 2010). Researchers employ standardized conditions, mechanical aids, representative samples, and error-correction strategies to ensure precise data collection. However, a drawback of this approach is that observed individuals may not exhibit natural behaviors in the controlled environment. For instance, it can be used to assess the impact of play therapy on children, comparing an experimental group that receives therapy in a controlled setting to a non-controlled group without therapy. On the other hand, non-controlled observation takes place in natural settings, devoid of precision instruments and stringent conditions. Researchers employing this method seek to understand behaviors and phenomena in their organic environments without external interference (Faird, 2022). For example, a researcher interested in studying the living conditions of people in extreme poverty within a rural community would conduct non-controlled observations by visiting the area. Here, standardized conditions and behavioral control are not applied, allowing for a more authentic understanding of the subjects' lives and behaviors.

These various types of observations enable researchers to choose the most suitable approach based on their research objectives, the level of interaction desired, and the need for standardized data collection.

4.2.3 When Do We Use the Observation Method for Data Collection?

The observation method is used for data collection in various research scenarios, particularly when researchers aim to gather firsthand information about behaviors, interactions, and events in their natural context. Here are common situations in which the observation method is employed:

Behavioral Research: Researchers use observations to study human or animal behaviors. For example, psychologists might observe children's play behavior to understand cognitive development, or ethologists might observe animals in their natural habitats to learn about their behaviors.

Social Sciences: In sociology and anthropology, observations are employed to study social interactions, cultural practices, and community dynamics. Observations can reveal insights into group behavior, norms, and societal structures.

Market Research: Businesses use observations to gather data on customer behavior and preferences. This includes observing how customers shop in stores, use products, or interact with websites and mobile apps.

Educational Research: Observations are common in educational settings to study classroom dynamics, teaching methods, and student engagement. Researchers might observe teachers' instructional strategies or students' learning behaviors.

Healthcare: Clinical observations are crucial in healthcare to assess patient symptoms, monitor vital signs, and track the progress of medical treatments. Observations are also used to assess the quality of healthcare services.

Environmental Science: Ecologists and environmental scientists use observations to study ecosystems, wildlife behavior, and natural phenomena. Field observations provide valuable data for ecological studies.

User Experience (UX) Research: In the design of products and interfaces, researchers observe how users interact with prototypes or existing systems. This helps improve usability and user satisfaction.

Criminal Justice: Observations are used by law enforcement agencies for surveillance and evidence gathering. Witness observations are crucial in criminal investigations.

Quality Control: Observations are employed in manufacturing and quality control processes to ensure products meet specific standards and specifications.

Ethnographic Research: Ethnographers immerse themselves in a particular culture or community, conducting participant observations to understand cultural practices, rituals, and social norms.

Naturalistic Studies: Researchers conducting naturalistic studies observe subjects in their natural environments without intervening or manipulating variables. This approach allows for ecologically valid data.

Longitudinal Studies: Observations are valuable for long-term studies that track changes over time, such as studies of child development or the impact of public policies.

In these and other research contexts, the observation method offers the advantage of collecting data in real-world settings, providing insights into natural behaviors and interactions. Researchers can choose between participant observations (where they interact with the subjects) and non-participant observations (where they remain unobtrusive observers) based on the research objectives and ethical considerations.

4.2.4 Conducting Observations

Indeed, conducting observations is a systematic process that involves several essential stages to ensure data collection is rigorous and meaningful. Here's an expanded view of these stages:

1. Planning:
   Define Research Objectives: Clearly outline the research goals and what you aim to observe. Determine the specific behaviors, events, or phenomena of interest.
   Select Observation Type: Decide on the type of observation you'll use, such as participant or non-participant, structured or unstructured, naturalistic or controlled.
   Context Identification: Identify the context or environment where the observations will occur. This could be a classroom, workplace, public space, or any setting relevant to your research.
2. Participant Selection:
   Relevance to Objectives: Select participants based on their relevance to the research objectives. Consider factors like demographics, roles, or characteristics that align with your research focus.

Informed Consent: If the observation involves human participants, obtain informed consent, ensuring they understand the purpose and scope of the observation and their role in it.

3. Data Collection:
   Systematic Observation: Conduct observations systematically, following the predetermined plan. Pay close attention to the behaviors, interactions, or events you intend to capture.
   Recording Tools: Utilize appropriate recording tools, which can include field notes, audio recorders, video cameras, or specialized software for digital observations.
   Observer Position: Decide on your position as an observer. Depending on the research goals, you may choose to be a visible observer (participants are aware of the observation) or a hidden observer (participants are unaware).

4. Recording:
   Note-Taking: If using field notes, ensure they are detailed, objective, and time-stamped. Record what you see, hear, and, if relevant, the context in which the observations occur.
   Audio and Video: If using audio or video recording, ensure the equipment is functioning properly and captures clear, unobtrusive data. Consider ethical and privacy considerations when recording people.

5. Transcription and Analysis:
   Transcription: If applicable, transcribe audio or video recordings into text. This step is crucial for further analysis.
   Data Coding: Analyze the observations by coding the data. Identify patterns, behaviors, or themes that emerge from the observations.
   Interpretation: Interpret the coded data to draw meaningful conclusions. Consider how the observed behaviors relate to your research objectives and what insights they provide.

Throughout the observation process, it's essential to maintain objectivity, minimize researcher bias, and ensure that the data collection aligns with the research objectives. Additionally, ethical considerations, such as informed consent and privacy, should be rigorously upheld when conducting observations involving human subjects.

4.2.5 Popularity of Observations in Research

Observations indeed hold a prominent place in research, especially in qualitative and field studies, owing to their unique advantages and popularity:

Rich Contextual Data: Observations excel in providing rich, contextual data. Researchers can witness behaviors, interactions, and events as they naturally occur, offering an in-depth understanding of the research subject within its real-life context.

Holistic Insights: Unlike some other data collection methods that focus on specific aspects or responses, observations aim to capture the entirety of a situation. This holistic perspective is valuable for exploring complex phenomena or multifaceted behaviors.

Minimized Recall Bias: Observations minimize recall bias, which can occur when participants are asked to recall past events or actions. By recording events in real-time, observations enhance the accuracy and reliability of data.

Nonverbal Communication: Observations allow researchers to capture nonverbal communication, such as body language, gestures, and facial expressions, which often convey nuances and emotions that verbal communication may not fully express.

Versatility: Observations can be adapted to various research settings and objectives. They are used in fields as diverse as anthropology, psychology, education, and market research.

Naturalistic Settings: For researchers interested in studying behavior in natural settings, observations are a preferred choice. This naturalistic approach ensures that participants act in their usual environment, minimizing the impact of artificial or controlled conditions.

Exploratory Research: Observations are well-suited for exploratory research, where the aim is to gain an initial understanding of a phenomenon, generate hypotheses, or identify areas for further investigation.

Participant Observation: In participant observation studies, where researchers become part of the observed group, a deep level of understanding can be achieved. This method is particularly valuable in ethnographic research.

Triangulation: Observations can be combined with other data collection methods, such as interviews or surveys, to enhance the validity and reliability of findings through triangulation.

Engagement with Participants: Observations often involve active engagement with participants or subjects, allowing researchers to establish rapport and gain insights that might not be as readily shared in other settings.

Despite their popularity, observations also come with challenges, such as the potential for observer bias, the need for careful planning, and the resource-intensive nature of data collection and analysis. Researchers must address these challenges to ensure the quality and rigor of their observational research.

4.2.6 How Do We Record and Store Observation Data?

Recording and storing observation data is crucial to maintaining the integrity and usefulness of your research. Here's how to effectively record and store observation data:

Recording Observation Data:

Field Notes: Field notes are a primary means of recording observation data. Use a notebook or digital device to write down detailed descriptions of what you observe. Include the date, time, location, and any relevant contextual information.

Photographs and Videos: If appropriate and allowed, use a camera or smartphone to take photographs or videos of the observed phenomena. Ensure that you have consent if people are identifiable in the media.

Audio Recordings: Audio recordings can capture conversations, sounds, or interviews related to the observation. Use a digital voice recorder or smartphone app for this purpose.

Checklists or Coding Sheets: For structured observations, you may use checklists or coding sheets to systematically record specific behaviors or events. This approach ensures consistency in data collection.

Storing Observation Data:

Digital Storage: Store digital observation data (e.g., photographs, audio recordings) on a secure computer, external hard drive, or server. Create dedicated folders or directories for different observations or research projects.

Cloud Storage: Utilize cloud storage services (e.g., Google Drive, Dropbox) for convenient backup, sharing, and access from multiple devices. Ensure the chosen service complies with data protection regulations.

File Naming and Organization: Adopt a clear and consistent file naming convention that includes relevant information like date, location, and a brief description. Organize files into folders or categories for easy retrieval.

Backup: Regularly back up your observation data to prevent loss due to technical failures or data corruption. Use multiple backup methods, such as external drives and cloud storage.

Data Security: Protect sensitive observation data with strong passwords and access controls. Limit access to authorized personnel only. Comply with data protection laws and ethical guidelines.

Metadata: Maintain metadata alongside your observation data, including observation logs, consent forms, and any contextual information. Metadata provides essential context and aids in data management.

Data Retention Policy: Develop a data retention policy that outlines how long you will retain the data after the research project concludes. Ensure it complies with ethical and legal requirements.

Data Encryption: If your observation data contains sensitive information, consider encrypting it. Encryption tools or software can safeguard data both in storage and during transmission.

Data Analysis and Reporting: Prepare your observation data for analysis by organizing and labeling it appropriately. Use data analysis software or tools if applicable.

Data Access and Sharing: If you plan to share or publish your observation data, follow ethical protocols and obtain the necessary permissions and consent. Anonymize or de-identify data to protect privacy when needed.

Effective recording and storage of observation data ensure that it remains accessible, secure, and usable for analysis and reporting, contributing to the success of your research project.

4.2.7 Merits of Observations

Conducting observations as a data collection method offers several merits that make it a valuable approach in research:

Rich and Detailed Data: Observations provide rich and detailed data about behaviors, interactions, and events as they naturally occur. This level of detail is often unattainable through other methods.

Contextual Understanding: Observations allow researchers to understand phenomena within their natural context. This contextual insight is particularly beneficial for studying complex or multifaceted behaviors.

Minimized Recall Bias: Observations minimize the risk of recall bias that can occur when participants are asked to remember and report past behaviors or events. Researchers witness actions as they happen, enhancing data accuracy.

Nonverbal Communication: Observations capture nonverbal communication, including body language, facial expressions, and gestures, providing valuable insights into emotions, attitudes, and unspoken messages.

Holistic Perspective: Observations aim to capture the entire situation or setting, providing a holistic perspective that considers all relevant elements and factors.

Naturalistic Settings: When researchers conduct observations in natural settings, participants behave in their usual environment, reducing the impact of artificial conditions or researcher-induced changes.

Exploratory Research: Observations are particularly useful for exploratory research where the primary goal is to gain a preliminary understanding of a phenomenon, generate hypotheses, or identify areas for further investigation.

Triangulation: Observations can be combined with other data collection methods, such as interviews or surveys, to enhance data validity and reliability through triangulation.

Participant Observation: In participant observation studies, where researchers become part of the observed group, an insider's perspective is gained, fostering deeper understanding and trust.

Real-Time Data: Observations provide real-time data, enabling researchers to capture dynamic processes and changes as they unfold.

Engagement with Participants: In some forms of observations, researchers actively engage with participants, which can help establish rapport and gain insights that might not be as readily shared in other settings.

Unobtrusive Method: Observations can be unobtrusive, meaning that participants are less likely to modify their behavior because they are being observed, as they might in a structured interview.

Objective Data: Observations aim to be objective, as researchers record what they see without relying on participants' interpretations or memories.

While observations offer numerous merits, they also come with challenges, such as the potential for observer bias, the need for careful planning and training, and the resource-intensive nature of data collection and analysis. Researchers must be aware of these challenges and take steps to address them to ensure the validity and reliability of their observational research.

4.2.8 Demerits of Observations

Despite their many merits, observations as a data collection method also have some inherent limitations and demerits:

Observer Bias: One of the most significant challenges is observer bias, where the researcher's presence and perspective can influence what they observe and record. Researchers may inadvertently focus on certain aspects, miss others, or interpret behaviors subjectively.

Limited Generalizability: Observations often provide context-specific data. Findings from observations in one setting may not generalize to other contexts or populations. This limits the external validity of the research.

Time-Consuming: Observations can be time-consuming, especially if they involve prolonged periods of data collection. This can strain research resources, particularly if multiple observers are needed.

Resource-Intensive: Conducting observations, especially in naturalistic settings, can require significant resources, including time, personnel, and equipment. This can be a barrier for researchers with limited resources.

Ethical Concerns: Observing individuals without their knowledge or consent in public settings raises ethical concerns related to privacy. In research involving sensitive or vulnerable populations, obtaining informed consent can be challenging.

Subjectivity in Interpretation: While observations aim to be objective, the interpretation of observed behaviors is inherently subjective. Different observers may interpret the same behavior differently, leading to variability in data analysis.

Intrusiveness: Even when participants are aware of being observed, the presence of an observer can alter their behavior, a phenomenon known as the Hawthorne effect. This can compromise the naturalness of the data.

Limited Insight into Mental Processes: Observations primarily capture observable behaviors and interactions. They may provide limited insight into participants' thoughts, emotions, or internal mental processes.

Selective Attention: Observers may unintentionally focus on behaviors or aspects of the situation that align with their preconceived notions or expectations, potentially overlooking important details.

Difficulty in Recording Complex Interactions: In settings with multiple simultaneous interactions or complex dynamics, it can be challenging for observers to capture and record all relevant information accurately.

Data Overload: Observations can generate vast amounts of data, which can be overwhelming to manage and analyze. Researchers must carefully plan data recording and storage.

Potential for Reactivity: In some cases, participants may alter their behavior due to the awareness of being observed, even if they are not explicitly informed about the research purpose. This can lead to inaccurate data.

To address these demerits, researchers must employ rigorous methodologies, minimize observer bias through training and inter-rater reliability checks, consider ethical implications, and carefully plan the scope and duration of observations. Combining observations with other data collection methods and using triangulation can also enhance the robustness of findings.

In summary, observations serve as a powerful tool for capturing genuine behaviors and interactions within real-life contexts. Through various observation types, researchers gain access to in-depth insights that contribute to a deeper understanding of human behavior and the intricacies of social dynamics. While observations provide contextually rich data, researchers must navigate challenges related to subjectivity, observer effect, and resource demands while reaping the benefits of this method's authenticity and validity.

4.3 In-depth Case Interviews

An in-depth interview stands as one of the primary and widely utilized methods for collecting information from participants in qualitative research (Azam, 2022). It is characterized by its flexibility and lack of strict structure, allowing for open dialog and exploration within the confines of the research objectives. In this approach, both the interviewer and respondent engage in a free-flowing conversation, enabling a deeper understanding of the participant's perspectives, experiences, and insights without rigid constraints. In-depth case interviews are a cornerstone of qualitative research, offering a rich and nuanced exploration of individual experiences, perspectives, and contexts. These interviews provide a platform for in-depth conversations that delve into the intricacies of a specific case or phenomenon, allowing researchers to uncover hidden insights and uncover the underlying complexities. Through carefully crafted questions and active listening, in-depth case interviews offer a window into the unique narratives and lived realities of research participants.

4.3.1 Definition of In-depth Case Interviews

An in-depth interview can be described as a dialog between an interviewer and a respondent, with its primary focus on the respondent's self-perception, life experiences, and viewpoints, all conveyed in their own words. It serves as a qualitative research technique characterized by the undertaking of detailed one-on-one interviews with a small number of participants, intending to delve into their unique perspectives regarding a specific subject, event, or circumstance (Azam, 2022). In-depth case interviews are qualitative research methods characterized by in-depth, semi-structured, or open-ended conversations between a researcher and a participant or a group of participants. These interviews aim to explore and understand a specific case, individual, or phenomenon comprehensively. Researchers typically employ probing questions and active listening techniques to encourage participants to share detailed narratives, personal experiences, and perspectives, allowing for a thorough investigation of the subject under study. In-depth case interviews are particularly valuable for gaining insights into complex, context-specific issues and providing a deeper understanding of the case's unique dynamics.

4.3.2 Types of In-depth Case Interviews

In-depth case interviews can take several forms, each suited to different research objectives and contexts. Some common types of in-depth case interviews include:

Standardized, Open-Ended Interviews: These interviews involve a set of predetermined open-ended questions that allow participants to provide detailed responses. The questions are designed to gather comprehensive information about the case under study.

Semi-Structured Interviews: Semi-structured interviews use a flexible interview guide with key topics and questions. While the researcher has a general structure to follow, they can adapt and probe deeper based on the participant's responses. This approach allows for both structure and flexibility.

Unstructured Interviews: Unstructured interviews are highly flexible and open-ended. They lack a predefined set of questions, giving participants the freedom to share their experiences and perspectives in their own words. Researchers guide the conversation based on the participant's narrative.

Narrative Interviews: Narrative interviews focus on collecting detailed life stories or personal narratives from participants. Researchers encourage participants to recount their experiences chronologically, emphasizing the narrative aspect of the interview.

Key Informant Interviews: Key informant interviews involve speaking with individuals who have specialized knowledge or expertise related to the case. These interviews aim to gather insights from those with insider information or unique perspectives.

Expert Interviews: Expert interviews are conducted with individuals recognized as authorities in a specific field or subject matter. These interviews seek in-depth insights and specialized knowledge related to the case.

Longitudinal Interviews: Longitudinal interviews are conducted over an extended period, allowing researchers to track changes and developments in the case over time. They provide a dynamic view of the subject.

Life History Interviews: Life history interviews delve into a participant's entire life story, exploring their experiences, milestones, and influences. These interviews are often used in biographical or autobiographical research.

The choice of interview type depends on the research objectives, the nature of the case, and the depth of understanding required. Researchers select the most appropriate approach to collect rich, context-specific data that aligns with their study's goals.

4.3.3 Rapport in In-depth Case Interviews

Rapport in in-depth case interviews is a fundamental aspect of the qualitative research process. It entails the establishment of a strong and positive interpersonal connection between the interviewer and the interviewee. This connection is pivotal as it sets the tone for the entire interview and significantly influences the depth and authenticity of the data collected.

Building rapport in an in-depth case interview involves creating an environment where the interviewee feels at ease, respected, and valued. It is about fostering trust and openness, which, in turn, encourages the interviewee to share their perspectives, experiences, and insights openly and honestly. Rapport building is not a one-size-fits-all approach; it varies depending on the context, the individuals involved, and the nature of the research.

Effective rapport building often includes active listening, where the interviewer pays close attention to what the interviewee is saying without interrupting or judging. It also involves demonstrating empathy and showing a genuine interest in the interviewee's point of view. A respectful and non-judgmental attitude is essential to create an atmosphere where the interviewee feels safe to express themselves.

Furthermore, rapport extends beyond the interview itself. It starts with the initial contact and continues throughout the research process. Researchers should be mindful of their communication, both verbal and non-verbal, as any signs of insincerity or indifference can hinder rapport building. When rapport is successfully established, it can yield several benefits. Interviewees are more likely to provide detailed and insightful responses, delve into their experiences, and share nuanced perspectives. This, in turn, enriches the data collected and enhances the overall quality of the research findings. In essence, rapport building is a foundational step in ensuring that in-depth case interviews produce valuable and authentic insights.

4.3.4 When Do We Use In-depth Case Interviews for Data Collection?

In-depth case interviews are used for data collection in various research contexts when researchers aim to gather detailed, nuanced information about a specific case or phenomenon. Here are some scenarios when in-depth case interviews are commonly employed:

Exploratory Research: When researchers seek to explore and understand a new or poorly understood phenomenon, in-depth case interviews help generate hypotheses and initial insights.

Complex Cases: In cases where the subject matter is intricate, multi-faceted, or involves numerous variables, in-depth case interviews provide a means to dive deep into each aspect.

Qualitative Research: In-depth case interviews are a primary data collection method in qualitative research. They are used to capture the experiences, perspectives, and narratives of participants.

Longitudinal Studies: In-depth interviews conducted over time (longitudinal interviews) are valuable for tracking changes, developments, and trajectories in individuals' lives or organizations.

Policy and Program Evaluation: Researchers use in-depth case interviews to assess the impact of policies, interventions, or programs on individuals or communities, capturing their experiences and feedback.

Human Behavior Studies: When studying human behavior, motivations, beliefs, or attitudes, in-depth interviews offer a means to explore the underlying reasons and contexts.

Biographical or Life History Research: In studies involving individuals' life stories, influences, and trajectories, life history interviews capture detailed narratives.

Complex Problem Solving: In professional or organizational contexts, in-depth interviews are used to understand complex issues, challenges, and decision-making processes.

Cultural and Sociological Research: In-depth interviews are valuable for examining cultural norms, values, and practices, as well as sociological phenomena within specific contexts.

Expert Knowledge: Researchers often conduct in-depth interviews with experts in a particular field to gain specialized insights and knowledge.

In-Depth Exploration: When the depth of understanding is paramount and requires participants to elaborate on their experiences, emotions, or perceptions, in-depth case interviews are the method of choice.

In summary, in-depth case interviews are employed when researchers aim to delve deeply into a specific case or phenomenon, seeking rich, context-specific data that provides a comprehensive and detailed understanding. These interviews are particularly valuable in qualitative research and contexts where in-depth exploration and narrative insights are essential.

4.3.5 How to Conduct In-depth Case Interviews?

Conducting in-depth case interviews is a structured yet flexible process that involves careful planning, effective communication, and skillful interviewing. Here are the key steps to conduct in-depth case interviews:

Define Research Objectives: Clearly outline the research objectives and the specific information you aim to gather through the interviews. Ensure that the objectives align with your research questions and overall study goals.

Select Participants: Identify and recruit participants who have direct experience or knowledge related to the case or phenomenon of interest. Consider the principle of purposive sampling to select individuals who can provide rich, informative insights.

Develop an Interview Guide: Create a semi-structured interview guide that outlines the topics, themes, and open-ended questions you intend to cover during the interviews. The guide serves as a flexible framework to ensure consistency while allowing for spontaneous exploration.

Obtain Informed Consent: Before conducting interviews, obtain informed consent from participants. Explain the purpose of the interview, the voluntary nature of participation, and how their data will be used and anonymized, ensuring ethical considerations are met.

Choose the Interview Setting: Select an appropriate and comfortable location for the interviews, ensuring privacy and minimal distractions. Depending on the context, interviews can be conducted face-to-face, over the phone, or via video conferencing.

Conduct the Interviews: During the interviews, establish rapport with participants by creating a welcoming and non-judgmental atmosphere. Begin with introductory questions to put participants at ease. Follow the interview guide while allowing flexibility for follow-up questions and probing to encourage in-depth responses.

Active Listening: Practice active listening throughout the interview. Pay close attention to what participants say, their tone, body language, and emotions conveyed. Use non-verbal cues such as nodding to indicate understanding and empathy.

Probing and Clarification: Ask probing questions to delve deeper into responses and clarify any ambiguities. Encourage participants to provide detailed accounts and elaborate on their experiences, perspectives, or decisions.

Record the Interviews: Use a reliable recording method, such as audio recording or note-taking, to capture the interview content accurately. Ensure participants are comfortable with the chosen recording method and obtain their consent.

Transcribe and Analyze: Transcribe the interview recordings if necessary. Then, analyze the interview data by identifying themes, patterns, and key insights. Consider using qualitative analysis software for systematic coding and categorization.

Maintain Confidentiality: Safeguard the confidentiality and anonymity of participants by securely storing interview data and using pseudonyms or codes in reporting.

Member Checking (Optional): Depending on your research design, you may conduct member checking, where participants review and verify the accuracy of interview transcripts or findings (for detail, please see Creswell & Creswell, 2017).

Report Findings: Present your findings in a clear and organized manner, using quotes and examples from the interviews to support your conclusions. Ensure that your reporting respects participants' anonymity and privacy.

Ethical Considerations: Continuously adhere to ethical guidelines throughout the process, including obtaining consent, protecting participants' rights, and addressing any ethical dilemmas that may arise.

Reflection and Iteration: Reflect on your interviewing process, including strengths and areas for improvement. Consider conducting follow-up interviews or refining your interview guide if necessary.

Effective in-depth case interviews require both technical skills in conducting interviews and a deep understanding of the research context.

Practice, training, and feedback can help researchers enhance their interviewing skills over time.

4.3.6 How Do We Record and Store In-depth Case Interviews Data?

Recording and storing in-depth case interview data is a crucial part of the research process to ensure the accuracy, confidentiality, and accessibility of the collected information. Here's a step-by-step guide on how to record and store in-depth case interview data:

Recording In-depth Case Interviews:

Select a Recording Method: Choose a recording method that best suits your research and the preferences of your participants. Common methods include audio recording, video recording, or written notes. Ensure that the chosen method respects participants' comfort and privacy.

Test Equipment: If using audio or video recording, test your recording equipment (microphones, cameras, software, etc.) before the interviews to ensure it functions properly. Check for background noise and clarity.

Seek Participant Consent: Obtain informed consent from participants to record the interviews. Clearly explain the recording process, how the data will be used, and the measures in place to protect their identity.

Initiate Recording: Start recording at the beginning of each interview. Ensure that the recording device is functioning correctly and capturing clear audio or video throughout the session.

Back-Up Recording: Consider having a backup recording method or device in case of technical issues with the primary recording. This helps prevent data loss.

Monitor the Interview: If you are conducting face-to-face interviews, pay attention to non-verbal cues, body language, and participant reactions, which may not be captured by audio or video alone.

Document the Interview Context: Make notes during the interview regarding any contextual information, observations, or impressions that may be relevant during the analysis phase.

Storing In-depth Case Interview Data:

Transcription (if applicable): If you recorded audio or video interviews, transcribe the content verbatim into text format. This step is essential for qualitative analysis. You can use transcription software or hire professional transcription services.

Organize Data: Create a structured and consistent system for organizing your interview data. This may include naming files logically, labeling transcripts, and maintaining a clear folder structure.

Data Security: Store interview data securely to protect participants' confidentiality. Use encryption for digital files and password protection for access. Keep physical copies, if any, in a locked and controlled environment.

Back-Up Data: Regularly back up your interview data to prevent loss due to technical failures or data corruption. Consider using cloud storage or external hard drives for secure backup.

Data Retention and Destruction: Adhere to ethical guidelines and legal requirements regarding data retention and destruction. Determine how long you need to retain the data and securely dispose of it when it's no longer needed.

Access Control: Limit access to interview data only to authorized researchers and team members who require it for analysis. Maintain an access log to track who accesses the data and when.

Anonymize Data: Before sharing or reporting interview data, ensure that it is anonymized to protect participants' identities. Replace real names with pseudonyms and remove any personally identifiable information.

Documentation: Maintain clear and detailed documentation of your data management processes, including recording methods, file locations, and access protocols. This documentation aids in transparency and reproducibility.

Ethical Review: If your research involves sensitive or personal information, seek ethical approval and guidance from your institution's ethics committee regarding data storage and protection.

Collaboration and Sharing: If collaborating with other researchers, establish clear protocols for sharing and storing data securely, adhering to data protection regulations.

By following these steps for recording and storing in-depth case interview data, you can ensure the integrity of your research while safeguarding the privacy and confidentiality of your participants.

4.3.7 Merits of In-depth Case Interviews

In-depth case interviews are a valuable research method that offers several merits for data collection and analysis. Here are some key advantages of using in-depth case interviews:

Rich and Detailed Data: In-depth case interviews provide rich, detailed, and contextually embedded data. Researchers can explore complex issues thoroughly, allowing for a deep understanding of the research topic.

Flexibility: This method is flexible and adaptable to various research objectives and settings. Researchers can tailor the interview guide to suit the specific case and research questions, making it suitable for diverse studies.

Exploratory Research: In-depth case interviews are particularly well-suited for exploratory research where little prior information is available. They allow researchers to generate hypotheses, identify patterns, and develop theories from scratch.

Participant Perspectives: Interviews capture the perspectives, experiences, and voices of participants directly. This method is valuable for understanding how individuals perceive and make sense of their experiences.

Contextual Insights: Researchers can gain insights into the broader context in which the case operates. This includes organizational dynamics, cultural factors, and situational influences, which are crucial for comprehensive analysis.

Participant Empowerment: In-depth interviews can empower participants by giving them a platform to express their views, concerns, and experiences. This can be especially important in studies involving marginalized or underrepresented groups.

Qualitative Data: In-depth interviews generate qualitative data that can be used for thematic analysis, content analysis, and other qualitative research methods. This type of data is valuable for exploring complex social phenomena.

Trust and Rapport: When conducted effectively, in-depth interviews build trust and rapport between the researcher and the participant. This can lead to more open and honest responses, enhancing data quality.

Clarification and Probing: Researchers can ask follow-up questions and seek clarification during interviews, ensuring a deeper understanding of participant responses. This probing ability helps uncover hidden meanings.

Theory Development: In-depth case interviews often contribute to theory development. Through the analysis of multiple cases, researchers can identify patterns and construct theoretical frameworks.

In-Depth Comparison: Researchers can conduct comparative analyzes across cases, which can lead to broader insights and generalizations about the research topic. This is particularly useful in multiple-case study designs.

Holistic Understanding: In-depth case interviews allow for a holistic understanding of a case or phenomenon, taking into account various dimensions and perspectives, that quantitative methods alone may not capture.

Despite these merits, it's essential to recognize that in-depth case interviews also have limitations and may not be suitable for all research questions or situations. Researchers should carefully consider their research objectives and the specific context when choosing this method.

4.3.8 Demerits of In-depth Case Interviews

While in-depth case interviews offer numerous advantages for data collection, they also come with certain limitations and potential drawbacks. Here are some of the demerits associated with in-depth case interviews:

Subjectivity: In-depth case interviews are inherently subjective, as they rely on the perspectives and interpretations of both the interviewer and the interviewee. This subjectivity can introduce bias into the data.

Time-Consuming: Conducting, transcribing, and analyzing in-depth interviews can be time-consuming, especially when dealing with multiple cases or lengthy interviews. This can limit the scalability of this data collection method.

Resource-Intensive: In-depth interviews require resources, including personnel, time, and often financial resources for participant recruitment and compensation. This can be a barrier for researchers with limited budgets.

Sample Size: In-depth case interviews typically involve a small sample size due to their intensive nature. As a result, findings may not be easily generalizable to larger populations.

Potential for Social Desirability Bias: Participants may provide responses that they believe align with societal norms or what they perceive as socially acceptable. This can lead to social desirability bias, where participants provide answers that present them in a favorable light.

Interviewer Bias: The interviewer's presence and interactions can influence participant responses. Interviewer bias, where the interviewer inadvertently shapes or directs responses, is a concern that needs to be managed.

Difficulty in Standardization: Achieving consistency and standardization in in-depth interviews can be challenging, particularly when multiple interviewers are involved or when the research spans across different cases or settings.

Limited Quantitative Data: In-depth interviews primarily yield qualitative data. Researchers seeking quantitative data for statistical analysis may need to use complementary methods or incorporate structured questions within interviews.

Limited Cross-Case Comparisons: While in-depth interviews provide rich insights into individual cases, they may not facilitate easy cross-case comparisons. This limitation can hinder the generalizability of findings.

Dependence on Interviewee's Recall: Participants are asked to recall past experiences, events, or behaviors. Memory limitations or biases in recall can affect the accuracy of data.

Ethical Considerations: Ethical challenges can arise, particularly when dealing with sensitive or personal topics. Researchers must ensure participant confidentiality and informed consent.

Interpretation Complexity: Analyzing qualitative data from in-depth interviews can be complex, involving the identification of themes, patterns, and underlying meanings. Interpretation may be influenced by the researcher's perspective.

Limited Objectivity: Achieving objectivity in the analysis of qualitative data can be difficult, and interpretations may be influenced by the researcher's preconceived notions or biases.

Lack of Control Over Environment: In-depth interviews conducted outside controlled settings may be influenced by external factors, such as noise, interruptions, or distractions.

To mitigate these limitations, researchers often use a mixed-methods approach, combining in-depth interviews with quantitative data collection methods to enhance the rigor and comprehensiveness of their research. Careful planning, training of interviewers, and transparency in reporting findings can also help address some of these challenges.

4.4 Focus Group Discussions (FGDs)

Focus Group Discussions (FGDs) are a dynamic data collection method that brings together a small group of participants (generally 6–12) to engage in guided discussions on a specific topic. This method encourages participants to share their perspectives, opinions, and experiences while interacting with one another. FGDs offer a unique platform to explore collective viewpoints, generate insights through group dynamics, and uncover shared attitudes. This section explores the intricacies of FGDs as a data collection method, covering their definition, types, the process of conducting FGDs, the role of notetaking, as well as their merits and demerits.

4.4.1 Definition of Focus Group Discussions (FGDs)

FGDs involve interactive group discussions led by a researcher or facilitator, focusing on a specific research topic. a focus group represents an informal conversation involving a select group of participants centered around a specific topic, as outlined by Wilkinson (2004). It constitutes a qualitative research method where a typically homogenous group of individuals is invited to express their perceptions, opinions, beliefs, and attitudes regarding various subjects, ranging from products and services to concepts, advertisements, ideas, or packaging (Khan & Abedin, 2022). A hallmark of focus groups is the assembly of individuals who share common interests or characteristics, brought together by a moderator. This moderator harnesses the group's dynamics, interactions among members, and exchanges with participants to extract valuable insights about the particular issue under consideration. Participants share their thoughts, experiences, and opinions within the group setting, allowing for the exploration of diverse perspectives and the emergence of group dynamics.

4.4.2 Types of FGDs

Focused Group Discussions (FGDs) can be classified based on various criteria, such as participants and objectives. Here's an expanded explanation of the types of FGDs:

Homogeneous FGDs: In homogeneous FGDs, participants share similar characteristics, such as demographics (age, gender, occupation) or experiences (e.g., members of a specific profession or patients with a particular medical condition). The homogeneity of the group allows for in-depth exploration of common viewpoints, attitudes, or experiences within that specific subgroup. This type of FGD is particularly useful when researchers want to understand the perspectives of a particular demographic or subgroup.

Heterogeneous FGDs: In contrast to homogeneous FGDs, heterogeneous FGDs involve participants with diverse characteristics. These groups are intentionally composed of individuals with different backgrounds, experiences, or viewpoints. Heterogeneous FGDs are valuable when researchers aim to explore a wide range of perspectives, capture contrasting viewpoints, or encourage creative thinking through the interaction of diverse participants. These discussions can be especially useful for brainstorming sessions or when the research objective is to identify potential solutions to complex problems.

Expert and Non-Expert FGDs: Depending on the research context, FGDs can also be categorized as expert and non-expert groups. Expert FGDs include individuals with specialized knowledge or experience in the topic under discussion, such as professionals, academics, or policymakers. Non-expert FGDs involve participants who may not have specialized knowledge but can provide valuable insights based on their personal experiences or general perspectives. This distinction is useful when researchers want to compare expert opinions with those of the general population or specific target groups.

Single-Topic and Multi-Topic FGDs: FGDs can focus on a single topic or cover multiple topics within the same discussion. Single-topic FGDs delve deeply into one specific subject, allowing participants to explore it comprehensively. Multi-topic FGDs, on the other hand, address several related or unrelated subjects during the same discussion. Researchers choose between these types based on the research objectives and the need for depth or breadth of discussion.

Exploratory and Confirmatory FGDs: FGDs can also be classified as exploratory or confirmatory based on their research goals. Exploratory FGDs are conducted when researchers seek to gain initial insights, generate hypotheses, or explore a new research area. Confirmatory FGDs, conversely, are used to validate or confirm existing research findings or hypotheses. These categories help researchers decide whether FGDs should serve as a starting point for research or as a means of corroborating existing knowledge.

Traditional and Online FGDs: With advancements in technology, FGDs can be conducted both in traditional face-to-face settings and online platforms. Traditional FGDs involve participants physically gathering in a specified location for discussions. Online FGDs, on the other hand, are conducted remotely through digital communication tools and platforms. Online FGDs offer flexibility in terms of participant location but may present challenges related to technology and non-verbal communication.

The choice of FGD type depends on the specific research objectives, the characteristics of the target population, and the depth and breadth of information researchers aim to gather during the discussion. Researchers should carefully consider these factors when designing and conducting FGDs for their studies.

4.4.3 When Do We Use FGDs for Data Collection?

FGDs are particularly useful for data collection in several situations:

Exploring Complex Topics: FGDs are well-suited for exploring complex and multifaceted topics where participants' interactions can shed light on diverse perspectives, attitudes, and experiences. They are valuable when you want to delve deep into a subject.

Understanding Group Dynamics: FGDs excel at capturing group dynamics, making them useful for studying how individuals interact, influence each other, and collectively construct meaning around a topic.

Generating Ideas and Insights: FGDs are effective for brainstorming and idea generation. They can help generate new concepts, hypotheses, or solutions through group interactions and discussions.

Exploring Diverse Perspectives: When you want to understand a topic from various angles or gather diverse viewpoints, FGDs allow participants to share their experiences, beliefs, and opinions openly.

Pretesting Materials: FGDs can be used to pretest research instruments, questionnaires, educational materials, or interventions. Participants' feedback and discussions can help refine these materials before broader deployment.

Assessing Attitudes and Perceptions: FGDs are valuable for assessing attitudes, perceptions, and social norms within a group. They can uncover shared beliefs and societal influences on individual perspectives.

Evaluating Programs or Interventions: FGDs can be employed to assess the impact of programs, interventions, or policies within a target community or population. They provide insights into how participants perceive and experience these initiatives.

Market Research: In marketing and product development, FGDs are commonly used to understand consumer preferences, opinions, and behaviors related to products, services, or brands.

Policy Development: FGDs can inform the development of policies and regulations by gathering input from stakeholders and affected parties. They facilitate dialog and consensus-building.

Cultural Studies: FGDs are valuable in cultural studies to explore cultural practices, beliefs, and traditions within a community or group.

Healthcare and Medical Research: In healthcare, FGDs can be employed to investigate patient experiences, healthcare access, health behaviors, or perceptions of medical treatments.

Education: FGDs are used in educational research to explore student experiences, teaching methods, and learning environments.

Community Engagement: FGDs can engage communities in decision-making processes, allowing residents to voice their concerns and preferences regarding local issues.

Conflict Resolution: FGDs are sometimes used in conflict resolution and peacebuilding efforts to bring together stakeholders with differing viewpoints to discuss and seek common ground.

It's important to note that FGDs require careful planning, skilled facilitation, and thoughtful analysis to be effective. The choice to use FGDs should align with the research objectives and the nature of the data you aim to collect. Additionally, considerations such as participant selection, setting, and ethical concerns should be taken into account when employing this data collection method.

4.4.4 Conducting Focus Group Discussions

Conducting Focus Group Discussions (FGDs) is a multi-stage process that requires careful planning and facilitation. Here's an expanded explanation of the stages involved in conducting FGDs:

Planning:

- Define Research Objectives: Clearly outline the research objectives and what you aim to achieve through the FGDs. Determine the specific information or insights you seek.
- Select FGD Type: Choose the type of FGD that aligns with your research goals. Decide whether a homogeneous or heterogeneous group would be more appropriate.
- Participant Recruitment: Develop a recruitment strategy to identify and invite suitable participants. Consider factors like demographics, experience, and availability.

- Schedule and Logistics: Determine the date, time, and location for the FGDs. Ensure that the venue is comfortable and conducive to open discussions.

Participant Selection:

- Criteria: Select participants based on predetermined criteria that align with your research objectives. Ensure diversity or homogeneity, depending on your FGD type.
- Informed Consent: Obtain informed consent from participants, explaining the purpose of the FGD, the expected duration, and the use of recordings. Address any questions or concerns.

Facilitation:

- Skilled Facilitator: Assign a skilled facilitator who is experienced in guiding group discussions. The facilitator should remain neutral, encourage participation, and manage the flow of conversation.
- Group Dynamics: Be mindful of group dynamics. Encourage quieter participants to share their thoughts, and manage dominant voices to ensure everyone has an opportunity to contribute.
- Icebreakers: Use icebreakers or warm-up activities to help participants become comfortable and build rapport with one another.

Discussion Guide:

- Structured Guide: Develop a structured discussion guide that outlines the topics, questions, and prompts to be covered during the FGD. The guide provides a framework for the discussion while allowing flexibility for spontaneous insights.
- Opening and Closing: Start with an introduction, objectives, and ground rules. Conclude by summarizing key points discussed and thanking participants for their input.

Recording:

- Audio/Video Recording: Record the FGDs using audio or video recording equipment. Ensure that participants are aware of the recording, and respect their privacy and confidentiality.
- Note-Taking: In addition to recordings, consider having a note-taker to capture non-verbal cues, group dynamics, and key observations during the discussion.

Transcription and Analysis:

- Transcription: Transcribe the recorded FGDs into text format. This textual data will be the basis for analysis.
- Data Analysis: Analyze the transcribed data to identify themes, patterns, and insights. Use qualitative analysis techniques such as thematic analysis or content analysis to derive meaningful findings.

Throughout the FGD process, it's essential to maintain a respectful and open atmosphere where participants feel comfortable expressing their opinions. Researchers should also consider the cultural and ethical aspects relevant to the group being studied. The insights gathered from FGDs can provide valuable qualitative data for research, offering a deeper understanding of participants' perspectives, attitudes, and experiences.

4.4.5 How Do We Record and Store FGD Data?

Recording and storing FGD data is essential to ensure the information is organized, secure, and accessible for analysis. Here's how to effectively record and store FGD data:

Recording FGD Data:

Audio Recording: FGD sessions are often audio recorded using digital voice recorders or smartphones with recording apps. Ensure participants' consent for audio recording at the beginning of each session.

Video Recording: Video recording can capture not only audio but also non-verbal cues, facial expressions, and group dynamics. Obtain informed consent for video recording when applicable.

Note-Taking: In addition to audio or video recording, facilitators and note-takers should capture key points, non-verbal cues, and any relevant observations during the FGD. Note-taking can help in understanding context and nuances.

Transcription: Transcribe the recorded FGD sessions, converting spoken words into written text. Transcriptions can be done manually or with the assistance of transcription software or services.

Storing FGD Data:

Digital Storage: Store FGD data digitally in a secure and organized manner. Create a dedicated folder or directory on your computer or a secure server for FGD data. Use strong passwords for access.

Cloud Storage: Consider using cloud storage services like Google Drive, Dropbox, or Microsoft OneDrive for backup and easy access to your data from anywhere with an internet connection. Ensure that the cloud storage provider complies with data protection regulations.

File Naming and Organization: Adopt clear and consistent file naming conventions, such as including the FGD session date, location, and a unique identifier. Organize files into subfolders or categories to group related data.

Backup: Regularly back up your FGD data to prevent loss due to technical issues or data corruption. Use external hard drives, additional cloud backups, or backup software for redundancy.

Data Security: Protect sensitive FGD data with strong passwords and access controls. Limit access to authorized team members or collaborators. Be aware of data protection regulations and ensure compliance.

Data Encryption: If your data contains sensitive or personal information, consider encrypting it. Encryption tools or software can protect data both in storage and during transmission.

Metadata: Maintain metadata, such as FGD session logs, consent forms, and data documentation, alongside the FGD data. Metadata provides context and helps you keep track of the sessions.

Data Retention: Develop a data retention policy specifying how long you will retain the data after the research project concludes. Ensure compliance with ethical and legal requirements.

Data Access and Sharing: If you plan to share or publish your FGD data, follow appropriate protocols and obtain informed consent from participants. Anonymize or de-identify data when necessary to protect privacy.

Data Management Software: Consider using data management software or qualitative analysis tools designed for FGD data. These tools can assist in organizing, coding, and analyzing your data efficiently.

By following these guidelines for recording and storing FGD data, you can maintain data integrity, ensure confidentiality, and facilitate data analysis and reporting for your research project.

4.4.6 Note Taking in FGDs

Note-taking plays a crucial role in FGDs as it complements audio or video recordings and provides valuable context and insights. Here's an expanded explanation of the importance of note-taking in FGDs:

Capture Key Points: Notes help in recording essential information, such as participants' responses to specific questions, significant comments, or unexpected insights. They serve as a quick reference for important findings during data analysis.

Contextual Details: Notes can provide contextual details about the discussion, including the sequence of topics, transitions between subjects, and any noteworthy occurrences or events during the FGD.

Non-Verbal Cues: While recordings capture verbal communication, note-taking allows the researcher to document non-verbal cues, such as body language, facial expressions, and gestures. These cues can reveal emotions, hesitations, or group dynamics that might not be evident in the spoken words alone.

Facilitator Observations: The facilitator's observations and reflections can be noted during the FGD. These observations might include the facilitator's assessment of group dynamics, the level of engagement among participants, or any challenges faced during the discussion.

Real-Time Insights: Note-taking provides an opportunity for the researcher to jot down real-time impressions, questions, or thoughts that arise during the FGD. These notes can help in generating follow-up questions or areas for deeper exploration.

Enhanced Analysis: When it comes to analyzing FGD data, notes can be invaluable. Researchers can refer to their notes to gain a deeper understanding of the discussion context, participant dynamics, and any nuances that might impact the interpretation of findings.

Quality Assurance: Having detailed notes acts as a form of quality assurance. In case of any issues with audio or video recordings, well-maintained notes can serve as a backup source of data.

To maximize the effectiveness of note-taking in FGDs, the note-taker needs to be attentive, organized, and skilled in summarizing key points. Clear and legible notes are more valuable during the data analysis phase. Researchers should also establish a system for labeling notes, such as indicating the speaker's initials or numbering significant statements for easy reference.

While audio or video recordings are primary data sources, the combination of recordings and comprehensive notes ensures a more comprehensive and nuanced understanding of the FGD, ultimately leading to richer research findings.

4.4.7 Merits of FGDs

FGDs offer several merits that make them a valuable data collection method in qualitative research. Here's an expanded explanation of the merits of FGDs:

Group Dynamics: FGDs excel in capturing the dynamics of group interactions. Participants engage in discussions, react to each other's viewpoints, and build on one another's ideas. This dynamic exchange of thoughts and opinions can reveal how group dynamics influence decision-making, consensus-building, or disagreement.

Diverse Perspectives: FGDs are particularly effective in eliciting diverse perspectives on a given topic. Participants often come from different backgrounds, experiences, or demographics, which can lead to a wide array of viewpoints and insights. This diversity enriches the data by providing a comprehensive understanding of the research topic from various angles.

Idea Generation: FGDs are conducive to idea generation and brainstorming. As participants engage in dialog, they may collectively generate new ideas, innovative solutions, or novel insights. The synergy created within the group can spark creative thinking and lead to the exploration of unexpected concepts or connections.

In-Depth Exploration: FGDs allow for in-depth exploration of participants' opinions, attitudes, and experiences. Researchers can delve into the underlying reasons behind participants' viewpoints by asking follow-up questions and encouraging elaboration. This depth of exploration is especially valuable for gaining a nuanced understanding of complex topics.

Rich Qualitative Data: FGDs yield rich qualitative data that provides context and depth to research findings. The narrative nature of FGDs allows participants to share personal stories, anecdotes, and real-life experiences, adding a human dimension to the data.

Peer Validation: Participants often validate or challenge each other's perspectives within the group setting. This peer validation can lead to a more accurate representation of shared beliefs, attitudes, or cultural norms. Researchers can observe how participants react to differing viewpoints and whether consensus or divergence emerges.

Efficiency: FGDs can be an efficient way to gather data from multiple participants simultaneously. Compared to conducting individual interviews, FGDs allow researchers to collect insights from several participants in a single session, making them a time-effective method for data collection.

Flexibility: FGDs offer flexibility in terms of adapting to the research objectives. Researchers can adjust the discussion guide or prompts in real-time based on the group's responses, allowing for a dynamic and responsive approach to data collection.

Participant Empowerment: FGDs empower participants by giving them a voice and a platform to express their opinions and experiences. This participatory approach can enhance participant engagement and satisfaction with the research process.

Cultural Insights: In cross-cultural research, FGDs can provide valuable insights into cultural norms, values, and behaviors. Participants' discussions may reveal cultural nuances and shed light on how culture shapes their perspectives.

Overall, FGDs are a versatile qualitative research method that excels in uncovering group dynamics, diverse perspectives, and in-depth insights, making them a valuable tool for researchers seeking to explore complex topics from multiple angles.

4.4.8 Demerits of FGDs

FGDs, while valuable, also come with their set of limitations and demerits. Here's an expanded explanation of the demerits of FGDs:

Dominant Voices: One of the common challenges in FGDs is the presence of dominant participants who may monopolize the discussion. These dominant voices can overshadow quieter participants, limiting the diversity of perspectives and potentially skewing the data towards one viewpoint.

Social Desirability Bias: FGD participants may be inclined to provide responses they believe are socially desirable or acceptable within the group context. This bias can lead to participants conforming to perceived norms, which may not accurately reflect their true opinions or experiences.

Groupthink: Group dynamics can sometimes lead to a phenomenon known as groupthink, where participants strive for consensus rather than expressing dissenting views. This can result in a lack of critical examination of ideas and a tendency to conform to the group's prevailing opinion.

Complex Analysis: Analyzing FGD data can be complex and time-consuming due to the multiple voices, interactions, and layers of information within the discussions. Researchers must sift through a wealth of qualitative data to identify themes, patterns, and meaningful insights.

Limited Generalizability: FGD findings are context-specific and may not be easily generalizable to broader populations. The insights gained from a specific group of participants may not apply to individuals outside of that context, making it important to consider the scope of generalization.

Moderator Influence: The effectiveness of FGDs heavily relies on the skills and neutrality of the moderator. A biased or inexperienced moderator can inadvertently influence the direction of the discussion, potentially introducing bias into the data.

Logistical Challenges: Organizing and conducting FGDs can be logistically challenging. Coordinating the schedules of multiple participants, securing a suitable venue, and managing audio or video recording equipment can be logistical hurdles.

Limited Privacy: FGDs involve participants discussing sensitive topics in a group setting. Some participants may feel uncomfortable sharing personal experiences or opinions in front of others, potentially leading to self-censorship or withholding of information.

Difficulty with Sensitive Topics: FGDs may not be suitable for discussing highly sensitive or taboo topics, as participants may be hesitant to openly share their thoughts or experiences in a group setting.

Small Group Size: The effectiveness of FGDs can be influenced by the size of the group. Very small groups may lack diversity of perspectives, while very large groups can be challenging to facilitate effectively.

Despite these demerits, FGDs remain a valuable qualitative research method when conducted with careful consideration of their limitations. Researchers should be aware of these potential pitfalls and take steps to mitigate them to ensure the quality and reliability of the data collected through FGDs.

In summary, Focus Group Discussions offer a valuable approach to gathering insights through interactive group interactions. This method allows researchers to explore the depth of collective perspectives, uncover group dynamics, and generate new insights through participant interactions. While FGDs provide a platform for diverse viewpoints and dynamic interactions, researchers must navigate challenges related to group dynamics and biases while harnessing the benefits of this method's depth and richness.

4.5 Key Informant Interviews (KIIs)

Key Informant Interviews (KIIs) are a qualitative data collection method that involves interviewing individuals who possess specialized knowledge or expertise about a particular topic. These individuals, known as key informants, provide valuable insights and perspectives based on their in-depth understanding of the subject matter. KIIs offer a focused and expert-driven approach to data collection, contributing to a comprehensive exploration of complex topics. This section delves into the intricacies of KIIs as a data collection method, encompassing their definition, types, the process of conducting KIIs, their applicability, as well as their merits and demerits.

4.5.1 Definition of Key Informant Interviews (KIIs)

KIIs involve interviews with individuals who are recognized as experts, authorities, or insiders within a specific field or subject. These individuals possess in-depth knowledge, experiences, and insights that make their perspectives valuable for understanding complex topics. A KII can be defined as a structured conversation facilitated by a skilled interviewer with an expert within a specific field or an influential figure within a community (Akhtar, 2022). The purpose of conducting a KII is to gather valuable insights and information from individuals who hold expertise, leadership roles, or a deep understanding of the subject matter. These interviews are typically aimed at eliciting recommendations and solutions based on their specialized knowledge and perspectives (Akhtar, 2022). Key informants may include community leaders, professionals, or residents who can provide valuable input and guidance relevant to the research or project at hand.

4.5.2 Types of KIIs

Key informant interviews can be conducted by face-to-face interviews and by telephone interviews. KIIs can be categorized based on the informants and objectives:

Expert Interviews: These interviews involve individuals who are acknowledged experts in a particular field or domain. They provide insights rooted in their extensive knowledge and experience.

Insider Interviews: Insider interviews involve individuals who have first-hand experience or unique perspectives due to their role, position, or affiliation within a specific context.

Key Stakeholder Interviews: These interviews are conducted with individuals who hold significant stakes or interests in a particular issue, project, or decision. They may include stakeholders from various sectors, such as government officials, community leaders, or industry representatives.

Policy Maker Interviews: Policy maker interviews involve individuals responsible for shaping policies, regulations, or laws related to the research topic. These interviews provide insights into the policy development process and its impact.

Frontline Worker Interviews: Interviews with frontline workers, such as healthcare professionals, educators, or social workers, can offer valuable insights into the challenges and opportunities they encounter in their daily roles.

Community Leader Interviews: Community leader interviews focus on individuals who hold leadership roles within a specific community or organization. They can provide perspectives on community dynamics, decision-making processes, and local issues.

Advocate or Activist Interviews: Interviews with advocates or activists involved in social, environmental, or political causes can shed light on the motivations, strategies, and challenges associated with advocacy work.

Consumer or User Interviews: These interviews involve individuals who are end-users or consumers of products, services, or systems. They provide feedback on their experiences, preferences, and needs.

Historical Interviews: Historical interviews are conducted with individuals who possess historical knowledge or have witnessed significant events relevant to the research topic. They help preserve oral histories and insights.

Cross-Cultural Interviews: In cross-cultural research, interviews may be conducted with individuals from different cultural backgrounds to explore cultural perspectives, practices, and experiences.

Interdisciplinary Interviews: Interdisciplinary interviews involve individuals with diverse expertise or backgrounds who can offer multifaceted insights into a complex issue. These interviews promote interdisciplinary collaboration and understanding.

Longitudinal Interviews: Longitudinal interviews are conducted with the same informants over an extended period to track changes, developments, or experiences over time.

Comparative Interviews: Comparative interviews involve interviewing informants from multiple locations, contexts, or groups to draw comparisons and contrasts, often used in cross-cultural or multi-site studies.

The choice of KII type depends on the research objectives and the specific insights needed. Additionally, the selection of key informants should align with the goals of the research and the expertise or perspectives required to address the research questions effectively.

4.5.3 Conducting Key Informant Interviews

KIIs are commonly employed in social research, and face-to-face KIIs hold distinct advantages. They foster a comfortable and congenial atmosphere for both the interviewer and the key informants (Akhtar, 2022). Conducted in a natural setting, face-to-face interviews provide the interviewer with a firsthand sense of the key informants' working environment or community context. This approach is well-suited for capturing non-verbal communication cues and allows for the convenience of asking follow-up and probing questions that naturally arise during free-flowing discussions. However, it's worth noting that in situations where there are constraints such as time limitations or the non-availability of high-profile key informants for face-to-face interviews, an alternative approach involving Internet-based interviews can also be considered (Akhtar, 2022).

Preparation: Before conducting KIIs, researchers should thoroughly prepare for the interviews. This includes defining clear research objectives, identifying potential key informants, and developing a comprehensive understanding of the research topic.

Ethical Considerations: Just like other data collection methods, ethical considerations are crucial. Researchers should obtain informed consent from key informants, ensuring they understand the purpose of the interview, the use of the data, and any potential risks or benefits.

Establishing Rapport: Building rapport with key informants is essential. Researchers should create a comfortable and respectful environment for the interviews, fostering open and honest communication.

Probing and Follow-up Questions: KIIs often involve probing and follow-up questions to delve deeper into key informants' insights. These questions can be tailored to the individual's expertise and responses.

Flexibility: While KIIs are guided by a structured interview guide, interviewers should be flexible to allow key informants to share their perspectives and insights in their own words. This flexibility can lead to unexpected but valuable insights.

Dealing with Sensitive Topics: In some cases, KIIs may touch on sensitive or personal topics. Interviewers should handle these discussions with sensitivity, respect, and empathy, ensuring key informants feel heard and valued.

Anonymity and Confidentiality: Researchers should discuss and respect key informants' preferences regarding anonymity and confidentiality, especially when dealing with sensitive information.

Quality Control: To ensure the quality of KIIs, interviewers should undergo training in interview techniques, active listening, and data capture. Peer debriefing and feedback can also enhance the quality of the interviews.

Data Management: Managing KII data, including recordings, transcripts, and notes, should be done securely and in compliance with data protection regulations.

Analysis: KII data analysis involves identifying themes, patterns, and expert insights. Researchers may use qualitative analysis techniques, such as thematic analysis or content analysis, to extract meaningful information from the interviews.

The successful conduct of KIIs relies on careful planning, effective communication, ethical considerations, and a focus on obtaining rich insights from knowledgeable individuals.

4.5.4 When Do We Conduct KIIs

Key Informant Interviews (KIIs) are typically conducted in research or investigative projects when you need in-depth information, insights, or expertise from individuals who possess specialized knowledge or experience in a particular field or domain. Here are common scenarios in which KIIs are conducted:

Expert Insights: When you require expert opinions or insights on a complex topic, particularly in fields like medicine, science, technology, or policy. Experts can provide deep knowledge and context.

Unique Perspectives: KIIs are useful when you need insights from individuals who have unique perspectives or experiences that are not readily available through other sources.

Validation or Verification: KIIs can be used to validate or verify information obtained from other sources, such as surveys or secondary data, by consulting experts or insiders who can confirm or elaborate on findings.

Complex Issues: In research involving complex or multifaceted issues, KIIs can help in unraveling intricate details and nuances that might not be apparent through other methods.

Policy Development: In the development of policies or guidelines, key informants, especially experts in the relevant field, can provide critical input to shape effective and evidence-based policies.

Program Evaluation: KIIs are conducted as part of program evaluations to assess the impact, effectiveness, or challenges of programs or interventions.

Exploratory Research: When researchers are exploring a new area or topic and need foundational knowledge to guide further investigation.

Qualitative Research: In qualitative research, KIIs can be used to gather qualitative data and narratives that help build a comprehensive understanding of a subject.

Case Studies: KIIs are often a component of case studies, where interviews with key stakeholders or experts provide insights into specific cases or scenarios.

Needs Assessment: KIIs can be used in needs assessments to understand the specific needs, preferences, or challenges faced by a particular group or community.

It's essential to carefully select key informants based on their expertise or unique perspectives relevant to your research objectives. Additionally, KIIs require thoughtful planning, ethical considerations, and skilled interviewers to ensure the quality and reliability of the data collected.

4.5.5 Field Setting and Suitability to Conduct a KII

The field setting for conducting a Key Informant Interview (KII) plays a crucial role in the quality of data collected. The suitability of the setting can impact the comfort and openness of the key informant, the ability to gather relevant information, and the overall success of the interview. Here are key considerations for selecting a suitable field setting for KIIs:

Natural Environment: Whenever possible, conduct KIIs in a natural environment familiar to the key informant. For example, if you're interviewing a community leader, interviewing at their office or within their community can be more comfortable and conducive to open dialog.

Privacy and Comfort: Ensure that the setting provides privacy and comfort to the key informant. This is essential for them to feel at ease and share information freely. If interviewing at their workplace or home, arrange for a quiet and confidential space.

Accessibility: Choose a location that is easily accessible to the key informant. This reduces any inconvenience and potential barriers to participation.

Cultural Sensitivity: Be culturally sensitive when selecting the field setting. Consider the cultural norms and preferences of the key informant and the community. Respecting their cultural context can foster rapport and cooperation.

Technology and Resources: Ensure that the chosen setting has the necessary technology and resources if you plan to use audio or video recording equipment. Test these tools beforehand to avoid technical issues during the interview.

Logistics: Plan the logistics of the interview, including transportation, refreshments, and any materials you may need. These details can contribute to a smooth and successful interview.

Safety and Security: Prioritize the safety and security of both the interviewer and the key informant. If the setting raises safety concerns, consider alternatives or take precautions.

Community Context: For interviews involving community leaders or residents, understanding the broader community context is important. This context can influence the dynamics of the interview and the responses provided.

Time Considerations: Be mindful of the key informant's schedule and availability. Schedule the interview at a time that is convenient and suitable for them.

Flexibility: Be adaptable. Sometimes, field settings may need to change due to unforeseen circumstances. Have contingency plans in place to address such situations.

In summary, the field setting for KIIs should prioritize the comfort, privacy, and cultural sensitivity of the key informant. A well-selected setting can enhance the quality of data collected and promote a positive interviewing experience.

4.5.6 Applicability of KIIs

Key Informant Interviews (KIIs) have specific applicability in research and investigative contexts where expert insights and nuanced perspectives are crucial. KIIs have been used in program evaluations and needs assessments (Lavrakas, 2008). Here are some key points that highlight the applicability of KIIs:

In-Depth Understanding: KIIs are valuable when you seek a profound understanding of a subject matter, especially when dealing with complex or specialized topics that require expert knowledge.

Expert Opinions: When you need expert opinions, advice, or recommendations, KIIs enable you to tap into the expertise of individuals who have a deep understanding of the field.

Validation and Verification: KIIs can be used to validate or verify information obtained through other means, such as surveys or secondary data sources. Experts can confirm or provide additional context to data.

Unique Perspectives: KIIs are applicable when you want to capture unique perspectives, experiences, or insights that are not readily available from the general population.

Policy Development: In the development of policies, guidelines, or best practices, KIIs help in shaping evidence-based recommendations and decisions.

Program Evaluation: KIIs are useful in evaluating the impact, effectiveness, or challenges of programs or interventions, especially when experts or key stakeholders are involved.

Emerging Fields: In emerging or rapidly evolving fields, KIIs can provide up-to-date information and insights that might not be available in existing literature.

Complex Issues: KIIs are applied when dealing with multifaceted or intricate issues where multiple perspectives and expert input are essential for a comprehensive analysis.

Qualitative Research: In qualitative research, KIIs serve as a primary method for gathering qualitative data, narratives, and detailed accounts.

Case Studies: KIIs are often used in case studies to obtain insights from key figures or participants in a specific case or scenario.

Program Planning: In the planning stages of programs or projects, KIIs help identify potential challenges, opportunities, and best practices.

Needs Assessment: KIIs are employed in needs assessments to understand the specific needs, preferences, and priorities of a particular group or community.

The applicability of KIIs lies in their capacity to provide nuanced, context-rich data that contributes to a more comprehensive and informed research outcome. Careful selection of key informants and a well-structured interview process are critical to their success in research projects.

4.5.7 How Do We Record and Store KIIs Data?

Recording and storing Key Informant Interviews (KIIs) data is a crucial aspect of the research process to ensure the information is accessible, secure, and organized. Here's how to record and store KII data effectively:

Recording KIIs Data:

Audio Recording: This is a common method for capturing KIIs data. You can use a digital voice recorder or a smartphone with a good-quality recording app. Ensure that the key informant consents to an audio recording before proceeding.

Video Recording: Video recording is suitable when you want to capture not only the audio but also non-verbal cues, facial expressions, and body language. Obtain informed consent for video recording as well.

Note-Taking: In some cases, you may rely on handwritten or typed notes to record the key informant's responses. Ensure that your note-taking is accurate and captures the essence of the conversation.

Transcription: If you use audio or video recording, you'll need to transcribe the interviews. Transcription involves converting spoken words into written text. This can be done manually or using transcription software and services.

Storing KIIs Data:

Digital Storage: Once you have recorded and transcribed the interviews, store the digital files securely. Create a dedicated folder or directory on your computer or a secure server for KIIs data. Encrypt sensitive files if necessary.

Cloud Storage: Consider using cloud storage services like Google Drive, Dropbox, or Microsoft OneDrive for backup and easy access to your data from anywhere with an internet connection. Ensure that the cloud storage provider complies with data protection regulations.

Naming and Organizing: Organize your data systematically. Use clear and consistent file naming conventions, such as including the interviewee's name, date, and a unique identifier. Create subfolders or categories to group related files.

Backup: Regularly back up your KIIs data to prevent loss due to technical issues or data corruption. Use external hard drives, additional cloud backups, or backup software for this purpose.

Data Security: Protect sensitive KII data with strong passwords and access controls. Limit access to authorized team members or collaborators. Be aware of data protection regulations in your jurisdiction and ensure compliance.

Data Encryption: If your data contains sensitive or personal information, consider encrypting it. Encryption tools or software can be used to protect data both in storage and during transmission.

Metadata: Maintain metadata, such as interview logs, consent forms, and data documentation, alongside the interview data. Metadata provides context and helps you keep track of the interviews.

Data Retention: Develop a data retention policy that outlines how long you will retain the data after the research project concludes. Ensure compliance with ethical and legal requirements.

Data Access and Sharing: If you plan to share or publish your KIIs data, follow appropriate protocols and obtain informed consent from participants. Anonymize or de-identify data when necessary to protect privacy.

Data Management Software: Consider using data management software or tools designed for qualitative data, which can help organize, code, and analyze your KIIs data more efficiently.

By recording and storing KII data following these guidelines, you can maintain data integrity, protect participant confidentiality, and facilitate data analysis and reporting for your research project.

4.5.8 Merits of KIIs

Expert Insights: KIIs provide access to expert knowledge, allowing researchers to tap into specialized perspectives and insights.

In-Depth Understanding: Key informants offer nuanced insights that can shed light on intricate aspects of the research topic.

Comprehensive Exploration: KIIs contribute to a comprehensive exploration of complex phenomena by capturing insights from those deeply involved.

Rich Contextual Information: KIIs provide rich contextual information that helps researchers understand the broader environment in which the research topic exists. Key informants can provide historical context, current trends, and future projections.

Credibility and Validation: Expert key informants lend credibility to the research findings. Their insights can validate or corroborate data obtained through other research methods, increasing the overall reliability of the study.

Tailored Responses: KIIs allow for tailored questioning and probing based on the key informant's responses. Researchers can adapt their questions in real-time to explore interesting or unexpected insights.

Real-World Applications: The insights gained from KIIs often have direct real-world applications. Policymakers, practitioners, and organizations can use this information to make informed decisions and improvements.

Holistic Perspectives: KIIs help in presenting a holistic view of the research topic by considering the perspectives of experts or key figures who have a comprehensive understanding of the subject matter.

Flexible and Adaptive: KIIs are flexible and can be adapted to various research objectives and topics. Researchers can select key informants from diverse backgrounds and fields.

Enhanced Research Quality: KIIs contribute to the overall quality of research by providing nuanced, expert-driven insights. They add depth to the data collected and strengthen the research's credibility.

Opportunity for Clarification: Researchers can seek clarification or elaboration on complex issues during KIIs, ensuring that they fully understand the key informant's viewpoints and insights.

Direct Engagement: KIIs involve direct engagement with key informants, fostering a deeper connection between the researcher and the participant. This can lead to more candid and detailed responses.

Unique Perspectives: KIIs often reveal unique, unconventional, or alternative viewpoints that may not be accessible through other data collection methods.

Timely Information: KIIs can provide timely information on rapidly evolving topics or emerging issues, as key informants are often at the forefront of their respective fields.

Exploring Controversial Topics: KIIs are suitable for exploring controversial or sensitive topics where expert opinions are essential for a balanced understanding.

Enhanced Data Validity: The data obtained from KIIs can enhance the validity of the research findings, especially in studies where expert insights are critical.

These merits highlight the significant advantages that Key Informant Interviews bring to the research process, making them a valuable tool for researchers aiming to gain expert perspectives and an in-depth understanding of their research topics.

4.5.9 Demerits of KIIs

Limited Generalizability: Findings from KIIs might not be easily generalizable due to the specificity of the participants' expertise.

Subjectivity: Key informants' opinions might be influenced by their personal biases or experiences.

Resource Intensive: Identifying and accessing key informants, as well as conducting interviews, can be resource-intensive.

Expert Availability: Experts might have busy schedules, making it challenging to secure their participation.

Limited Sample Size: KIIs typically involve a small number of key informants due to their specialized knowledge or expertise. This limited sample size may not capture the full diversity of perspectives on a given topic.

Response Bias: Key informants may provide responses that align with their professional or personal interests, potentially introducing bias into the data.

Expertise Variability: The quality and depth of insights obtained in KIIs can vary depending on the expertise and willingness of the key informants. Some may offer highly valuable insights, while others may provide less useful information.

Difficulty in Recruitment: Identifying and recruiting key informants can be challenging, especially if they are prominent figures in their field or have busy schedules.

Time-Consuming: KIIs can be time-consuming to arrange, conduct, and analyze. This can extend the overall timeline of the research project.

Subjective Analysis: Analyzing data from KIIs can be subjective, as it often involves interpreting nuanced qualitative information. This subjectivity may affect the reliability of findings.

Potential for Social Desirability Bias: Key informants may provide responses that they believe align with socially desirable norms or expectations, rather than expressing their genuine opinions.

Difficulty in Comparison: Comparing and synthesizing insights from multiple key informants can be challenging, especially if their viewpoints differ significantly.

Resource Demands: KIIs can be resource-intensive, requiring funding for participant compensation, transcription services, and other logistical needs.

Limited Representativeness: The perspectives of key informants may not fully represent the broader population or community, limiting the generalizability of findings.

Potential for Misrepresentation: If key informants are not entirely forthcoming or if their expertise is misrepresented, it can lead to inaccuracies in the data.

Ethical Considerations: Ensuring ethical conduct in KIIs, such as obtaining informed consent and addressing potential conflicts of interest, can be complex.

Difficulty in Validation: Validating the accuracy of information obtained in KIIs can be challenging, as it often relies on the credibility and trustworthiness of key informants.

These demerits highlight the important considerations and potential challenges associated with using Key Informant Interviews in research. Researchers should weigh these limitations against the benefits when deciding whether KIIs are an appropriate data collection method for their specific research objectives.

In summary, Key Informant Interviews offer a focused and expert-driven approach to data collection. By engaging individuals with specialized knowledge, researchers gain access to valuable insights and perspectives that contribute to a comprehensive understanding of complex topics. While KIIs offer expert insights and in-depth understanding, researchers must navigate challenges related to limited generalizability and resource demands while harnessing the benefits of this method's depth and expertise.

4.6 Community Mapping

Community Mapping is a participatory data collection method that engages community members in visually representing their environment, resources, and social dynamics. Through drawings, sketches, or maps, participants create a spatial representation of their community, highlighting key features, relationships, and assets. Community Mapping empowers participants to share their perspectives, contribute to decision-making, and provide insights into their local context. This section explores the intricacies of Community Mapping as a data collection method, encompassing its definition, types, process of conducting Community Mapping, its applicability, as well as its merits and demerits.

4.6.1 Definition of Community Mapping

Community Mapping involves the collaborative creation of maps, diagrams, or visual representations of a community's physical and social environment. Participants draw upon their knowledge and experiences to depict geographical features, landmarks, resources, relationships, and cultural elements that hold significance within their community.

4.6.2 Types of Community Mapping

Community mapping is a versatile data collection method that can be adapted to various research needs and contexts. Here are different types of community mapping approaches:

Physical or Geographic Mapping: This type of mapping involves creating visual representations of physical features within a community, such as streets, buildings, landmarks, and natural resources. It is commonly used in urban planning, disaster preparedness, and environmental studies.

Social Network Mapping: Social network mapping focuses on identifying and visualizing social connections and relationships within a community. Researchers map individuals or groups and their interactions, highlighting key influencers and information flow.

Resource Mapping: Resource mapping involves documenting the availability and distribution of resources within a community. This can include mapping healthcare facilities, schools, water sources, agricultural land, and other essential services and assets.

Participatory Mapping: In participatory mapping, community members actively engage in the mapping process. They contribute their knowledge and perceptions to create maps that reflect their perspectives on issues such as land use, cultural heritage, or local concerns.

Historical Mapping: Historical mapping involves tracing the historical development and changes in a community over time. It may include mapping changes in land use, demographic shifts, or cultural heritage sites.

Health Mapping: Health mapping focuses on the spatial distribution of health-related factors such as disease outbreaks, healthcare facilities, and environmental hazards. It can help identify health disparities and inform public health interventions.

Cultural Mapping: Cultural mapping aims to document and preserve cultural heritage, including traditions, languages, and historical sites within a community. It helps in safeguarding cultural diversity and promoting cultural tourism.

Asset Mapping: Asset mapping identifies and visualizes the assets and strengths of a community, including its human capital, community organizations, and economic opportunities. It is often used in community development and asset-based approaches.

Vulnerability Mapping: Vulnerability mapping assesses the susceptibility of a community to various risks, such as natural disasters or climate change impacts. It helps in disaster preparedness and risk reduction.

Urban Planning and Land Use Mapping: In urban planning, mapping is used to analyze land use patterns, transportation networks, and infrastructure development to inform urban design and policy decisions.

Environmental Mapping: Environmental mapping focuses on natural features and ecosystems within a community. It helps in conservation efforts, ecological studies, and environmental impact assessments.

Community Perception Mapping: This type of mapping captures community members' perceptions and subjective experiences related to their environment, safety, and quality of life.

Digital Mapping: With advancements in technology, digital mapping tools and Geographic Information Systems (GIS) are commonly used for creating interactive and layered maps that incorporate various data sources.

The choice of community mapping type depends on the research objectives, the information needed, and the preferences of the community and researchers. Often, a combination of mapping approaches may be used to create a comprehensive understanding of a community and its assets and challenges.

4.6.3 When Do We Conduct Community Mapping?

Community mapping is conducted in various contexts and for a range of purposes, depending on the specific research or project objectives. Here are some common situations when community mapping is conducted:

Urban and Regional Planning: Community mapping is frequently used in urban and regional planning to assess current land use, transportation infrastructure, housing distribution, and environmental resources. It helps urban planners make informed decisions about zoning, infrastructure development, and community services.

Disaster Preparedness and Response: Before and after natural disasters or emergencies, community mapping can be conducted to identify vulnerable areas, evacuation routes, emergency shelters, and critical resources. This information is vital for disaster management and response efforts.

Environmental Conservation: Community mapping is used to document and protect natural resources, biodiversity, and ecosystems. It helps in identifying areas of ecological significance and planning conservation initiatives.

Public Health: In public health research, community mapping is used to analyze the distribution of health services, disease outbreaks, and environmental health factors. It aids in understanding health disparities and planning healthcare interventions.

Community Development: Community-based organizations and development agencies use mapping to assess community assets, identify strengths, and plan for development projects. Asset mapping helps empower communities to leverage their resources.

Cultural Heritage Preservation: Cultural mapping is conducted to document and preserve cultural heritage, including historical sites, traditions, and languages. It plays a role in cultural preservation and tourism development.

Education Planning: In education, mapping is used to analyze the distribution of schools, educational resources, and student populations. It informs decisions about school placement and resource allocation.

Transportation and Mobility: Mapping is essential for transportation planning. It helps identify transportation needs, traffic patterns, and potential areas for infrastructure improvement.

Tourism Development: Tourism authorities and businesses use mapping to promote tourist destinations, highlight attractions, and provide visitors with information about local amenities and services.

Community Engagement: Participatory mapping engages community members in the process of documenting their local knowledge, concerns, and aspirations. It fosters community participation and ownership in decision-making.

Research and Data Collection: Researchers use community mapping to collect spatial data for a wide range of studies, from social science research to environmental studies and beyond.

Rural and Agricultural Development: Mapping is valuable in rural areas for assessing agricultural land, and water resources, and planning rural development initiatives.

Humanitarian Relief: During humanitarian crises, mapping helps aid organizations identify affected areas, plan relief operations, and assess the impact of disasters on communities.

Infrastructure and Utility Planning: Utilities such as water supply, electricity, and sanitation systems benefit from mapping to plan infrastructure development and maintenance.

Cross-Cultural Understanding: Cultural mapping contributes to cross-cultural understanding by documenting the cultural practices, traditions, and heritage of different communities.

The timing of community mapping can vary. It can be conducted as a one-time assessment or as an ongoing process to monitor changes and developments within a community. The key is to align the mapping effort with the specific goals and needs of the community or the research project.

4.6.4 Field Preparation for Community Mapping

Field preparation for community mapping is a crucial step to ensure the success and accuracy of the mapping process. It involves planning, gathering necessary resources, and engaging with the community. Here are the key steps in field preparation for community mapping:

Define Objectives and Scope:

- Clearly define the goals and objectives of the community mapping project. What do you aim to achieve through mapping?
- Determine the scope of the mapping effort, including the geographic area to be covered and the specific data to be collected.

Community Engagement:

- Engage with the community early in the process. Explain the purpose of the mapping project and seek their input and collaboration.

- Build trust and rapport with community members to ensure their active participation and willingness to share information.

Resource Gathering:

Identify the resources needed for mapping, including mapping tools (e.g., GPS devices, GIS software), stationery (e.g., maps, notebooks), and, if applicable, technology (e.g., smartphones or tablets).

Ensure you have an adequate supply of these resources for the fieldwork.

- Training and Capacity Building:
- Train mapping teams, including community members if they are involved in data collection, on how to use mapping tools and collect data accurately.
- Guide ethical considerations, such as respecting privacy and obtaining informed consent.

Data Collection Tools:

- Prepare data collection tools, such as questionnaires, survey forms, or mapping sheets, depending on the nature of the mapping project.
- Ensure that data collection tools are well-structured, easy to understand, and aligned with project objectives.

Logistics and Transportation:

- Arrange transportation to and within the mapping area, especially if it's a remote or hard-to-reach location.
- Consider logistical aspects such as accommodations, meals, and safety measures for the mapping team.

Permission and Access:

- Obtain any necessary permissions or permits to access private properties or restricted areas for mapping.
- If working in sensitive areas or with vulnerable populations, ensure that you have the required clearances.

Community Meetings and Workshops:

- Organize community meetings or workshops to explain the mapping process, answer questions, and gather initial data.
- Discuss the timeline, roles and responsibilities, and potential challenges.

Safety and Health Precautions:

- Prioritize safety by providing training on safe fieldwork practices, especially if mapping in rugged terrain, forests, or areas with potential hazards.
- Ensure that the mapping team has access to first-aid kits and emergency communication devices.

Documentation and Ethics:

- Document the mapping process, including who participated, key discussions, and any issues or challenges encountered.
- Reiterate the importance of ethical data collection, including respecting privacy and obtaining informed consent.

Quality Control:

Establish mechanisms for quality control, including data validation and verification processes.

Monitor the progress of mapping activities to ensure they align with project goals.

Community Mapping Plan:

- Develop a mapping plan that outlines the schedule, roles, and responsibilities of team members, including community participants.
- Include contingency plans for unexpected situations.

Logistical Support:

Arrange for any additional support or services needed in the field, such as interpreters or local guides.

Field preparation is a collaborative effort that involves effective communication with the community and meticulous planning to ensure that the community mapping project proceeds smoothly and achieves its objectives.

4.6.5 How to Conduct Community Mapping

Conducting community mapping involves a systematic process of data collection and mapping to create a visual representation of a community's assets, resources, challenges, and opportunities. Here's a step-by-step guide on how to conduct community mapping:

1. Define Objectives:
   Clearly define the objectives and goals of your community mapping project. What do you want to achieve, and what specific information are you seeking to collect?
2. Engage the Community:
   - Build strong relationships with community members. Explain the purpose and benefits of the mapping project, ensuring their active involvement and buy-in.
   - Collaborate with community leaders, organizations, or local authorities to gain support and access to resources.
3. Assemble a Mapping Team:
   Recruit and train a mapping team that can include community members, volunteers, or researchers. Ensure they understand the project's goals, ethical considerations, and data collection methods.
4. Collect Data:
   - Choose the data collection methods and tools that align with your project's objectives. Common methods include interviews, surveys, participatory workshops, or GPS technology.

- Collect data on various aspects of the community, such as infrastructure, natural resources, cultural landmarks, social services, and community assets.
- Encourage community members to share their knowledge, insights, and observations during data collection.

5. Organize and Record Data:
   - Organize and record the collected data systematically. This can include paper maps, digital maps, photographs, or written descriptions.
   - Use Geographic Information System (GIS) software or mapping applications to create digital maps if applicable.

6. Geographic Mapping:
   If using GIS or digital mapping tools, input the collected data to create maps. Ensure that the maps are visually appealing, easy to understand, and represent the community accurately.

7. Community Validation:
   - Share the preliminary maps and findings with the community for validation. Seek feedback and corrections to ensure accuracy.
   - Engage community members in discussions about the information presented on the maps.

8. Finalize Maps:
   - Incorporate community feedback and make necessary adjustments to the maps.
   - Produce final maps that depict the community's assets, resources, and challenges.

9. Share Findings:
   - Share the final maps and findings with the community members, stakeholders, and relevant organizations.
   - Use the maps to facilitate discussions and planning sessions with the community to address identified needs and opportunities.

10. Use for Decision-Making:
    - Encourage the community to use the maps as tools for decision-making, resource allocation, and community development initiatives.
    - Support community-driven projects that address the identified priorities.
11. Documentation:
    Document the entire community mapping process, including data collection methods, challenges faced, and lessons learned. This documentation can be valuable for future projects.
12. Maintain Ongoing Engagement:
    Maintain ongoing engagement with the community. Periodically update the maps to reflect changes and progress within the community.
13. Capacity Building:
    Consider conducting training sessions or workshops to build the capacity of community members in mapping techniques, enabling them to take ownership of future mapping efforts.

Community mapping is a collaborative and empowering process that not only helps collect valuable data but also strengthens community bonds and fosters local decision-making. It should be approached with respect for the community's knowledge and perspectives, emphasizing the importance of community ownership of the mapping process and its outcomes.

4.6.6 How Do We Record and Store Community Mapping Data?

Recording and storing community mapping data is essential to ensure the usefulness and longevity of the collected information. Here's how to effectively record and store community mapping data:

Recording Community Mapping Data:

Maps: Create detailed maps of the community, indicating relevant features, landmarks, and resources. You can use digital mapping software (e.g., GIS) or traditional paper maps.

Field Notes: Accompany your maps with detailed field notes that describe the community's characteristics, such as population, infrastructure, natural resources, and social dynamics. Include observations and interviews with community members.

Photographs and Videos: Take photographs or videos of key locations, buildings, infrastructure, and community activities. Ensure you have permission to capture images, especially if individuals are recognizable.

GPS Data: Use GPS devices or smartphone apps to record geographic coordinates of important points of interest. This data can be integrated with mapping software.

Community Interviews: Conduct interviews with community members to gather qualitative data about their perspectives, needs, and challenges. Record these interviews, with proper consent, and transcribe them for reference.

Community Workshops: If you involve community members in the mapping process through workshops or participatory activities, document the outcomes, discussions, and collaborative decisions.

Storing Community Mapping Data:

Digital Storage: Store digital mapping data, field notes, photographs, videos, and GPS data on secure computers, external hard drives, or servers. Ensure that your storage solutions are reliable and regularly backed up.

Cloud Storage: Utilize cloud storage services for convenient backup, sharing, and access from multiple devices. Be mindful of data security and privacy when using cloud storage.

File Naming and Organization: Adopt a consistent file naming convention that includes relevant information like location, date, and content. Organize files into folders or categories for easy retrieval.

Metadata: Create and maintain metadata records that provide context for the mapping data. Metadata should include information about data sources, collection methods, and any limitations.

Data Security: Protect sensitive community mapping data with strong passwords and access controls. Limit access to authorized personnel only and ensure compliance with data protection regulations.

Backup: Regularly back up your mapping data to prevent loss due to technical failures or data corruption. Use multiple backup methods, including external drives and cloud storage.

Data Sharing: If you plan to share community mapping data with others, ensure you have appropriate permissions from community members and follow ethical guidelines. Anonymize or de-identify data if necessary.

Data Retention Policy: Develop a data retention policy that specifies how long you will retain the data after the mapping project concludes. Ensure compliance with ethical and legal requirements.

Data Analysis and Reporting: Prepare your community mapping data for analysis by organizing and labeling it appropriately. Use mapping software and tools for analysis and visualization.

Effective recording and storage of community mapping data facilitate its accessibility and usability for future reference, research, planning, and community development initiatives. It also helps maintain the integrity of the collected information.

4.6.7 Applicability of Community Mapping

Community mapping can be applied in various contexts, including:

Urban Planning and Development: Community mapping helps urban planners identify infrastructure needs, land use patterns, and areas requiring development. It can support decisions related to transportation, housing, and public services.

Natural Resource Management: It aids in mapping natural resources, such as forests, rivers, and biodiversity. This information is essential for sustainable resource management and conservation efforts.

Healthcare Planning: Community maps can highlight healthcare facilities, disease hotspots, and areas lacking access to medical services. This data is valuable for healthcare planning and resource allocation.

Education: Mapping schools, educational facilities, and literacy rates can inform education policies and improve access to quality education.

Disaster Preparedness: Community mapping can identify vulnerable areas, evacuation routes, and emergency shelters, enhancing disaster preparedness and response efforts.

Cultural Preservation: It can help document cultural heritage sites, traditions, and indigenous knowledge, supporting cultural preservation initiatives.

Rural Development: In rural areas, community mapping can aid in agricultural planning, water resource management, and identifying opportunities for income generation.

Social Services: Identifying locations of social services like community centers, libraries, and youth clubs can help improve access to these services.

Community Empowerment: The process of community mapping empowers residents by involving them in decision-making and highlighting their contributions to the community.

Tourism: Mapping tourist attractions and amenities can boost the tourism industry and promote economic growth in the area.

Conflict Resolution: In conflict zones, community mapping can help identify sources of tension, resources that may be contested, and opportunities for conflict resolution.

Environmental Conservation: Mapping ecosystems and wildlife habitats aids in conservation efforts, ensuring the protection of biodiversity.

Transportation Planning: It can help design efficient transportation networks, including roads, public transit, and pedestrian pathways.

Community Organizing: Community maps can serve as tools for community organizers and advocacy groups to address issues such as housing, poverty, and social inequality.

Historical Research: Community mapping can be used in historical research to recreate past landscapes and understand historical events.

Community mapping is a versatile approach that adapts to the specific needs and priorities of a community. It promotes local participation, fosters a sense of ownership, and facilitates data-driven decision-making. However, it's essential to approach community mapping with sensitivity to cultural norms, privacy concerns, and ethical considerations.

4.6.8 Merits of Community Mapping

Participatory Approach: Community Mapping empowers participants to actively contribute their insights and knowledge. It fosters a sense of ownership among community members, making them central to the data collection and decision-making processes. This participatory approach not only enhances the accuracy and relevance of the maps but also strengthens community cohesion and collaboration.

Local Knowledge and Engagement: It leverages the knowledge and insights of community members, promoting active engagement and participation in decision-making processes.

Holistic Understanding: Community mapping provides a holistic view of a community's assets, resources, and challenges, enabling comprehensive planning and development.

Customization: It allows communities to tailor maps to their unique needs and priorities, ensuring that planning and development efforts are context-specific.

Spatial Awareness: Maps provide a clear visual representation of spatial relationships, helping community members, planners, and policymakers understand the geographic distribution of resources and services.

Improved Decision-Making: Decision-makers can use community maps to make informed choices about resource allocation, infrastructure development, and service provision.

Conflict Resolution: In conflict-prone areas, community mapping can help identify and address potential sources of tension and conflict, contributing to peacebuilding efforts.

Resource Management: It supports sustainable resource management by identifying natural resources, such as forests and water bodies, and promoting their responsible use.

Empowerment: Community members gain a sense of ownership and empowerment through the process of mapping their communities. This can lead to increased self-reliance and self-determination.

Advocacy and Awareness: Maps can be powerful advocacy tools, helping communities raise awareness of their needs and mobilize support from government agencies, NGOs, and donors.

Historical Documentation: Community mapping can serve as a historical record, documenting changes in land use, infrastructure, and demographics over time.

Partnerships: It encourages collaboration among stakeholders, fostering partnerships between community members, local authorities, and development organizations.

Transparency: Mapping processes are often transparent, ensuring that community members are well-informed about data collection, decision-making, and resource allocation.

Education and Capacity Building: Community mapping initiatives often involve training community members in data collection and mapping techniques, enhancing their skills and capacity.

Crisis Preparedness: Maps can be valuable tools in disaster risk reduction and emergency response, helping communities plan for and respond to crises.

Tourism and Economic Development: Community maps can promote tourism by highlighting attractions and amenities, boosting the local economy.

Cultural Preservation: Mapping cultural heritage sites and traditions contributes to cultural preservation efforts.

Environmental Conservation: It supports conservation efforts by identifying areas of ecological importance and potential threats to biodiversity.

Infrastructure Planning: Maps assist in planning infrastructure projects like roads, schools, healthcare facilities, and sanitation systems.

Improved Governance: Community mapping can lead to better governance by facilitating citizen engagement and accountability.

While community mapping offers numerous merits, it's essential to address potential challenges such as data privacy, the need for ongoing maintenance, and ensuring that marginalized voices are included in the process. Additionally, ethical considerations must be upheld throughout the mapping initiative.

4.6.9 Demerits of Community Mapping

Resource and Time Intensive: Community Mapping can be resource-intensive, requiring equipment, technology, and skilled personnel. The process can also be time-consuming, especially in larger or more complex communities.

Limited Technological Access: In some communities, access to mapping technology and tools may be limited, hindering the implementation of Community Mapping initiatives.

Data Accuracy: The accuracy of community-generated maps can vary depending on the participants' familiarity with mapping tools and their attention to detail. Errors or inconsistencies may arise if participants lack experience in mapping.

Subjectivity: Community Mapping may introduce subjectivity into the data collection process, as participants may prioritize certain features or locations based on their perspectives or interests.

Privacy Concerns: Mapping community details, especially in sensitive or vulnerable populations, can raise privacy concerns. Ensuring data protection and confidentiality becomes paramount.

Language and Cultural Barriers: In multicultural communities, language and cultural differences may complicate the mapping process, potentially leading to misinterpretations or misunderstandings.

Community Dynamics: Mapping can sometimes uncover underlying community tensions or conflicts, which may require careful management and resolution.

Maintenance Challenges: Once maps are created, there can be challenges in keeping them updated, especially if the community is dynamic or if external support diminishes over time.

Limited Generalizability: The insights and data gathered through Community Mapping are often specific to the mapped community and may not be easily generalizable to other contexts.

Technical Skills Requirement: Participants may need training to use mapping tools effectively, which can be a barrier in some cases.

It's essential to consider these demerits when planning and conducting Community Mapping initiatives and to mitigate potential challenges to ensure the success of the mapping project.

In summary, Community Mapping offers a participatory and visual approach to data collection, allowing community members to express their knowledge and perspectives through spatial representations. This method empowers participants, provides local insights, and captures cultural nuances. While Community Mapping enhances participatory engagement and local knowledge, researchers must navigate challenges related to accuracy, interpretation, and resource demands while harnessing the benefits of this method's community-driven insights.

4.7 Participatory Rural Appraisal (PRA)

Participatory Rural Appraisal (PRA) is a collaborative and participatory data collection method designed to engage community members in the assessment and analysis of their resources, needs, and priorities. PRA methodologies empower communities to actively contribute their knowledge, voice their concerns, and drive decision-making processes. This section delves into the intricacies of PRA as a data collection method, covering its definition, types, the process of conducting PRA, its applicability, as well as its merits and demerits.

4.7.1 Definition of Participatory Rural Appraisal (PRA)

Participatory Rural Appraisal (PRA) is a bottom-up approach to data collection that involves local communities in the process of identifying, analyzing, and addressing their issues and challenges. PRA methods emphasize collaboration, empowerment, and the inclusion of community perspectives. PRA is a dynamic approach to community-based research and development. It involves a range of participatory methods and tools that are conducted by local people themselves. These methods empower communities to collectively analyze their situation, identify challenges, and develop sustainable solutions (Chambers, 1994). PRA has been applied in a wide range of fields, including natural resources management (e.g., soil and water conservation, forestry, fisheries), agriculture, health, nutrition, food security, and poverty alleviation programs. Its strength lies in its ability to engage local communities in a collaborative and participatory process, leading to context-specific and sustainable solutions to various challenges.

4.7.2 Types of Participatory Rural Appraisal (PRA)

Participatory Rural Appraisal (PRA) encompasses various methods and techniques that facilitate community participation in the research and development process. Here are some common types of PRA methods:

Village Transect Walk: This method involves a guided walk through the community with community members. It helps in understanding the physical layout of the village, land use, and identifying significant landmarks.

Social Mapping: Participants create maps to represent social relationships, hierarchies, and networks within the community. This can help in identifying community leaders and influential figures.

Resource Mapping: Community members map natural resources, agricultural land, water sources, and other vital assets. This assists in identifying resource management issues and opportunities.

Seasonal Calendars: These calendars are created collaboratively with community members to depict seasonal variations, including planting and harvesting times, festivals, and resource availability.

Historical Timelines: Community members construct timelines to record historical events, changes, and significant developments within the community. This can provide valuable context for understanding current issues.

Community Meetings and Workshops: Conduct open meetings and workshops with community members to discuss various topics, gather information, and brainstorm solutions.

Transect Walks: Similar to the village transect walk, this method involves walking along a specific transect or linear path through the community or natural environment to gather information about environmental changes and land use.

Focus Group Discussions (FGDs): FGDs can also be considered a PRA method. These structured group discussions involve community members who share their experiences, perspectives, and ideas on specific topics.

Participatory Mapping: Community members create maps of their village or locality, highlighting significant features, boundaries, and resource locations. This method can be highly visual and informative.

Ranking and Scoring: Participants assign values or rank to various aspects or issues, helping to prioritize community concerns or preferences.

Problem Trees and Solution Trees: These visual tools help identify problems (in a problem tree) and potential solutions (in a solution tree) in a structured manner.

Matrix Ranking: Participants rank items or issues based on their significance or preference, helping to establish priorities.

Semi-Structured and Unstructured Interviews: Engaging in one-on-one or small group interviews with community members to gather in-depth qualitative information.

The choice of PRA methods depends on the research objectives, the community's preferences and capabilities, and the specific issues being addressed. PRA is flexible and adaptable, allowing researchers and development practitioners to select the most appropriate methods for a given context.

4.7.3 Guiding Principles of PRA

Participatory Rural Appraisal (PRA) is guided by several fundamental principles that underpin its approach to community engagement, data collection, and decision-making. These principles include:

Participation: PRA places a strong emphasis on active involvement and collaboration with community members. It recognizes that local people are the experts in their own lives and situations. They are not just sources of data but partners in the research process.

Empowerment: PRA seeks to empower communities by giving them the tools and knowledge to analyze and address their problems. It aims to shift the balance of power from external experts to local communities.

Holistic Approach: PRA takes a holistic view of communities and their challenges. It recognizes that social, economic, cultural, and environmental factors are interconnected and must be considered together.

Flexibility: PRA methods are flexible and adaptable to the specific context and needs of the community. There is no one-size-fits-all approach, and techniques are chosen based on what is most appropriate for the situation.

Ownership: PRA ensures that the results of the assessment belong to the community. The community members are the primary beneficiaries of the knowledge generated, and they have ownership of the process and outcomes.

Local Knowledge: PRA values and respects local knowledge and expertise. It recognizes that community members have a deep understanding of their environment and can provide insights that external experts may lack.

Non-Bias and Neutrality: PRA facilitators aim to remain neutral and non-biased, allowing communities to express their views and concerns freely. They do not impose their agendas or preconceived solutions.

Visual Tools: PRA often employs visual tools like maps, diagrams, and drawings to facilitate communication and understanding. These tools are especially useful when working with illiterate or non-literate communities.

Action-Oriented: PRA is not just about data collection; it is action-oriented. The ultimate goal is to use the information gathered to plan and implement community-driven solutions and improvements.

Transparency: PRA processes are transparent, and decisions are made collectively. This transparency helps build trust within the community and ensures that everyone has a voice in the process.

Ethical Considerations: PRA adheres to ethical principles, including informed consent, confidentiality, and respect for cultural norms and values. It ensures that the research process does not harm or exploit the community.

Sustainability: PRA aims to contribute to sustainable development. It seeks to identify solutions that are environmentally sustainable and socially equitable in the long term.

These guiding principles make PRA a powerful approach for understanding and addressing complex rural and community-based issues. They emphasize collaboration, local knowledge, and empowerment, ultimately leading to more effective and sustainable development outcomes.

4.7.4 Conducting Participatory Rural Appraisal (PRA)

In the context of Participatory Rural Appraisal (PRA), it's important to understand the process thoroughly.

Here's an expanded overview (Figure 4.1):

Entry and Rapport Building:

- Researchers initiate contact with the community, introduce themselves, and explain the purpose and goals of the PRA process.
- Building trust and rapport is essential. Researchers should respect local customs, engage community leaders, and establish good relationships with community members.
- Understanding the community's history, culture, and concerns is vital to build trust.

Method Selection:

- Collaborative decision-making is key. Researchers and community members together select the PRA methods to be used based on the research objectives and community context.
- Ensure that the selected methods are appropriate and culturally sensitive.

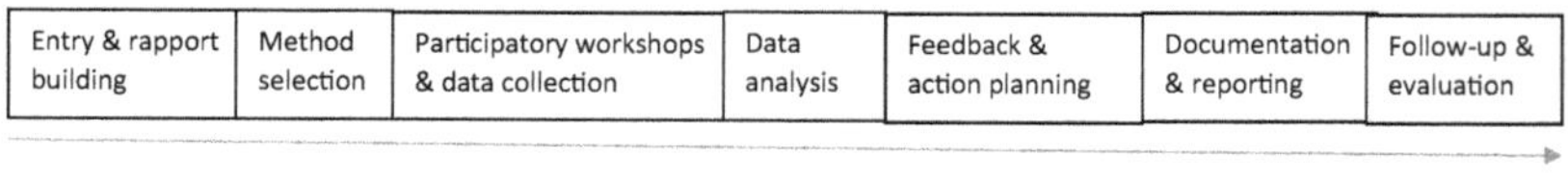

Figure 4.1. Steps to Conduct PRA.

Source: Author

Participatory Workshops and Data Collection:

- Organize participatory workshops or sessions within the community. These workshops are interactive and involve community members in data collection.
- Use selected PRA methods, such as social mapping, resource mapping, seasonal calendars, or historical timelines, to gather information. These methods should be carefully explained to participants.
- Facilitate discussions where community members actively share their knowledge, experiences, and insights.
- Encourage participants to use drawings, diagrams, or symbols to represent information visually.
- Capture both quantitative and qualitative data during these sessions.
- Ensure that everyone has an opportunity to contribute, and consider involving marginalized or less vocal community members.

Data Analysis:

- After data collection, work collaboratively with community members to analyze the gathered information.
- Identify patterns, trends, and key issues as a group.
- Community members can provide valuable insights during this analysis phase.
- Prioritize the identified issues based on the community's input and consensus.

Feedback and Action Planning:

- Share the results of the PRA with the community clearly and understandably.
- Engage community members in action planning to address the identified issues or to leverage opportunities.
- Ensure that the community takes ownership of the action plan, as this is crucial for the sustainability of any interventions or changes.

Documentation and Reporting:

- Document the entire PRA process, including methods used, data collected, analysis, and action plans.
- Prepare a comprehensive report that can be shared with stakeholders, including community members, local authorities, and organizations involved in the project.

Follow-Up and Evaluation:

- Periodically follow up with the community to assess the progress of action plans and address any emerging issues.
- Evaluate the impact of interventions or changes and make adjustments as necessary.

Throughout the PRA process, maintain a participatory and respectful approach, valuing local knowledge and expertise. Flexibility and adaptability are essential, as community dynamics and needs can evolve.

4.7.5 How Do We Record and Store PRA Data?

Recording and storing Participatory Rural Appraisal (PRA) data is crucial for preserving the insights and information gathered during the process. Here's a guide on how to effectively record and store PRA data:

Recording PRA Data:

Note-Taking: Assign a dedicated note-taker or scribe during PRA activities. This person should record key points, discussions, observations, and any visual representations (e.g., drawings, diagrams) made by participants.

Photographs and Videos: Capture visual evidence of PRA activities, such as photographs of maps, charts, and community discussions. Videos can also be used to document dynamic interactions.

Mapping and Diagrams: Create participatory maps, diagrams, and timelines during PRA exercises. These visual aids can convey valuable information about community resources, land use, and historical changes.

Transcriptions: If interviews or discussions take place, especially with community members, transcribe audio recordings into written form. Ensure that you have proper consent for recording.

Participant Feedback: Encourage participants to provide written or verbal feedback about their experiences with PRA activities, including what they found most valuable and any suggestions for improvement.

Group Outputs: Collect and preserve any outputs generated by the community during PRA sessions, such as community action plans, problem trees, or resource maps.

Storing PRA Data:

Digital Storage: Store digital data, including photographs, videos, and transcriptions, on secure computers, external hard drives, or cloud storage platforms. Use password protection and encryption for sensitive data.

Cloud Storage: Utilize cloud storage services to ensure data accessibility from multiple locations. However, be mindful of data security and privacy when using cloud storage.

File Naming and Organization: Develop a consistent file naming convention that includes relevant details such as location, date, and content. Organize files into folders or categories for easy retrieval.

Metadata: Create and maintain metadata records that provide context for the PRA data. Include information about the community, participants, facilitators, and data collection methods.

Data Security: Safeguard sensitive PRA data with strong passwords and access controls. Limit access to authorized personnel only and adhere to data protection regulations.

Backup: Regularly back up digital PRA data to prevent loss due to technical failures or data corruption. Use multiple backup methods, including external drives and cloud storage.

Data Sharing: If you plan to share PRA data with stakeholders or the community, ensure that you have appropriate permissions and follow ethical guidelines. Anonymize or de-identify data if necessary.

Data Retention Policy: Develop a data retention policy that outlines how long you will retain PRA data after the project concludes. Ensure compliance with ethical and legal requirements.

Data Analysis and Reporting: Prepare PRA data for analysis by organizing and labeling it appropriately. Use data analysis software or tools to derive insights and present findings.

Effective recording and storage of PRA data facilitate its accessibility and utility for future community development initiatives, project planning, and research. It also maintains the integrity of the knowledge and insights gained through the PRA process.

4.7.6 Applicability of Participatory Rural Appraisal (PRA)

Participatory Rural Appraisal (PRA) is applicable in various contexts where community involvement, local knowledge, and a deeper understanding of rural settings are essential. Here are some key contexts where PRA can be highly relevant:

Community Development: PRA is widely used in community development projects to involve local communities in identifying their needs, prioritizing development interventions, and planning and implementing solutions collaboratively.

Agriculture and Rural Livelihoods: PRA is valuable in agricultural projects to assess farming practices, crop choices, soil management, and the impact of climate change on rural livelihoods. It helps in designing sustainable and context-specific agricultural interventions.

Natural Resource Management: PRA is essential for the sustainable management of natural resources such as forests, water sources, and grazing lands. It allows communities to map resource locations, discuss usage norms, and plan conservation efforts.

Healthcare and Sanitation: PRA can be used to identify health challenges, sanitation issues, and healthcare access barriers in rural areas.

This information is crucial for designing effective health programs and interventions.

Education and Literacy: In the context of education, PRA can help assess educational infrastructure, identify barriers to access, and gather insights into local teaching and learning methods. This aids in tailoring education programs to local needs.

Social Inclusion and Gender Equality: PRA is instrumental in addressing issues related to social inclusion and gender equality by involving marginalized or vulnerable groups in discussions and decision-making processes.

Urban Planning and Infrastructure Development: While PRA is often associated with rural settings, it can also be applied in urban contexts for participatory urban planning, slum improvement, and the development of informal settlements.

Conflict Resolution and Peacebuilding: In areas affected by conflicts or tensions, PRA can be used to facilitate dialogs, build trust among conflicting parties, and gather insights into the root causes of conflicts to inform peacebuilding efforts.

Disaster Risk Reduction: PRA helps communities identify and assess local vulnerabilities to natural disasters, enabling them to develop disaster preparedness and mitigation strategies.

Monitoring and Evaluation: PRA methods are useful for ongoing monitoring and evaluation of development projects, allowing for adaptive management and adjustments based on real-time community feedback.

Research and Academic Studies: Researchers and academics often use PRA to gather qualitative data, especially when studying rural or marginalized communities, local knowledge systems, or traditional practices.

NGO and Development Projects: Non-governmental organizations (NGOs) and development agencies frequently employ PRA as a participatory and community-centered approach in their projects.

PRA's strength lies in its ability to empower communities, foster local ownership of development initiatives, and generate context-specific insights that lead to more effective and sustainable interventions. However, it's important to adapt PRA methods to suit the specific goals and cultural contexts of the communities being engaged.

4.7.7 Merits of Participatory Rural Appraisal (PRA)

Participatory Rural Appraisal (PRA) offers numerous merits and advantages, making it a valuable approach in various development and research contexts. Here are some of the key merits of PRA:

Community Empowerment: PRA empowers local communities by involving them directly in the research or development process. It recognizes community members as experts in their context and promotes active participation in decision-making.

Local Knowledge Utilization: PRA leverages the rich local knowledge of communities. By tapping into this knowledge, it ensures that interventions and solutions are contextually relevant and culturally sensitive.

Contextual Understanding: PRA provides a deeper and more nuanced understanding of the local context, including social structures, traditional practices, and community dynamics. This knowledge is essential for effective project design.

Holistic Perspective: PRA encourages a holistic view of issues by considering various aspects of community life simultaneously. It helps in identifying interconnections between different factors and their impact on development.

Cost-Effective: PRA often requires fewer resources compared to traditional research methods. It relies on existing community resources, such as local facilitators, and can be less expensive in terms of data collection.

Participatory Decision-Making: PRA facilitates participatory decision-making processes, which can lead to more sustainable and community-driven solutions. Communities have a say in setting priorities and designing interventions.

Rapid Data Collection: PRA can yield substantial information in a relatively short time. This is particularly useful for projects that require quick assessments or responses, such as disaster relief efforts.

Flexibility: PRA methods are adaptable to different contexts and issues. They can be customized to suit the specific needs and objectives of a project or research study.

Ownership of Outcomes: When communities actively participate in problem-solving and decision-making, they feel a sense of ownership over the outcomes. This often leads to better project sustainability.

Trust Building: PRA helps build trust between external organizations, researchers, and communities. It fosters positive relationships by respecting local knowledge and involving community members in every stage of the process.

Capacity Building: Through participation in PRA activities, community members can develop new skills, gain insights into their issues, and take on leadership roles in future initiatives.

Enhanced Impact: By tailoring interventions based on community insights, PRA can lead to more impactful and relevant development projects. This results in positive changes in the lives of community members.

Qualitative Data: PRA generates qualitative data that complements quantitative data. Qualitative insights help explain the "how" and "why" behind trends and statistics.

Cultural Sensitivity: PRA respects local cultures and traditions, reducing the risk of imposing solutions that may be culturally inappropriate or ineffective.

Adaptive Management: PRA allows for adaptive management, where project strategies can be adjusted based on real-time feedback from communities.

While PRA offers numerous advantages, it's important to recognize that its successful implementation requires skilled facilitators, culturally sensitive approaches, and a commitment to genuine participation. When conducted effectively, PRA can lead to more sustainable and impactful development outcomes.

4.7.8 Demerits of Participatory Rural Appraisal (PRA)

Participatory Rural Appraisal (PRA) is a valuable approach in many contexts, but it also comes with certain challenges and limitations that should be considered when planning its implementation. Here are some of the demerits of PRA:

Time-Consuming: PRA can be a time-intensive process. Building rapport with communities, conducting participatory activities, and reaching consensus on issues can take longer compared to traditional data collection methods.

Resource-Intensive: While PRA is often considered cost-effective, it still requires resources for training facilitators, organizing workshops, and conducting fieldwork. Adequate funding and logistical support are essential.

Facilitator Skills: Effective PRA relies on skilled facilitators who can guide community activities, manage group dynamics, and ensure that the process remains participatory. Finding and training qualified facilitators can be challenging.

Subjectivity: PRA activities are highly participatory and subjective by nature. While this can lead to rich insights, it may also introduce bias if not carefully managed. Facilitators and participants may have their perspectives and agendas.

Language and Communication: Language barriers and differences in communication styles can pose challenges in PRA. Ensuring that all participants can fully express themselves and understand each other is essential.

Limited Generalizability: PRA findings are context-specific and may not be easily generalizable to other settings or communities. The localized nature of PRA limits its external validity.

Complex Analysis: Analyzing PRA data can be complex due to the qualitative and participatory nature of the information. It may require specialized skills in qualitative data analysis.

Potential for Domination: In some cases, certain individuals or groups within the community may dominate discussions or decision-making processes, silencing others or skewing outcomes.

Resistance to Change: PRA can identify issues or challenges that require changes in existing practices or systems. Communities may be resistant to making these changes, especially if they are deeply rooted in tradition.

Limited Quantitative Data: PRA primarily generates qualitative data. While qualitative insights are valuable, they may not provide the statistical rigor required for certain types of research or decision-making.

External Influence: External organizations or researchers may inadvertently influence the outcomes of PRA processes. Communities might provide responses that they believe are expected by external actors, leading to social desirability bias.

Ethical Considerations: Ensuring ethical conduct in PRA, including informed consent and the protection of vulnerable populations, can be complex and requires careful attention.

Sustainability: PRA may not always lead to sustained community engagement or long-term project sustainability. Follow-up and ongoing support are often needed to maintain positive changes.

Despite these demerits, PRA remains a valuable tool for community-based research, development, and decision-making. Its strengths in promoting community participation, local knowledge utilization, and context-specific insights often outweigh these limitations when implemented thoughtfully and ethically.

In summary, Participatory Rural Appraisal is a collaborative and community-driven approach to data collection, placing local knowledge and community voices at the forefront. This method empowers communities to actively contribute to decision-making, address challenges, and shape their development. While PRA enhances community engagement and ownership, researchers must navigate challenges related to time, facilitation, and data interpretation while harnessing the benefits of this method's participatory insights and community-driven solutions.

4.8 Discourse Analysis

Discourse Analysis is a qualitative research method that examines language and communication to uncover underlying meanings, social constructs, and power dynamics. This method focuses on understanding how language shapes and reflects social realities, identities, and ideologies. Discourse Analysis allows researchers to explore how individuals and groups use language to convey their perspectives, beliefs, and social interactions. This section delves into the intricacies of Discourse Analysis as a data collection method, covering its definition, types, process of conducting Discourse Analysis, its applicability, as well as its merits and demerits.

4.8.1 Definition of Discourse Analysis

Discourse Analysis involves the systematic study of spoken, written, or visual language to unveil how language constructs meaning, influences perceptions, and shapes social interactions. The term "discourse" is employed by researchers in diverse ways to interpret language and textual elements, reflecting the evolving nature of communication due to technological advancements and the interdisciplinary aspects of discourse. Blommaert (2005) defines discourse as instances of communicative actions or meaningful symbolic behaviors, emphasizing the broader scope of communication beyond language alone. In contrast, Johnstone (2008) extends the notion of communication to encompass various forms of media, such as photography, clothing, music, architecture, and dance, all of which contribute to the creation of meaning.

This is a research approach that goes beyond simply studying language patterns within texts; it delves into the contextual aspects surrounding the use of language. This tool encompasses an examination of language in various forms, including written, spoken, sign language, or other symbolic communication processes like the use of metaphors, analogies, and symbolism to convey meaning (Sankar, 2022). As such, discourse analysis is concerned with both linguistic and non-linguistic elements, offering a comprehensive exploration of how communication functions within specific contexts and settings. This approach allows researchers to uncover the intricate ways in which language, along with its various nuances and forms, contributes to the construction of meaning in different social and cultural contexts. Michel Foucault (1972) introduces a different perspective, conceptualizing discourse as manifestations of larger systems of thought inherent to specific historical contexts. These systems shape the way certain concepts gain meaning in both thought and speech. Foucault argues that the ideologies preceding discourse, encompassing areas like religion, medicine, politics, power, and daily life, exert a significant influence on our thoughts and actions. These overarching systems that emerge within particular historical periods serve as guiding frameworks that elucidate and contextualize discourse. This tool serves a broader context of communication, examining both linguistic and non-linguistic elements.

4.8.2 Types of Discourse Analysis

Discourse Analysis comprises several distinct approaches, each uniquely suited to specific research goals and contextual considerations:

Critical Discourse Analysis (CDA): This approach delves into the examination of power dynamics, ideologies, and social disparities that are inherent in language usage. CDA seeks to uncover how language can perpetuate or challenge prevailing social norms and structures.

Multimodal Discourse Analysis: Unlike traditional textual analysis, this method extends its scope to include visual, auditory, and spatial elements within communication. It recognizes that meaning can be conveyed not only through words but also through images, sounds, and design elements.

Narrative Discourse Analysis: Focused on narratives and storytelling, this approach aims to understand how individuals construct their identities, convey their life experiences, and shape their perceptions of social realities through the stories they tell. It examines the structural and thematic elements of narratives to uncover deeper meanings.

These approaches to Discourse Analysis provide researchers with valuable tools to explore language and communication in diverse ways, offering insights into how discourse shapes and reflects social phenomena.

4.8.3 How Do We Conduct Discourse Analysis?

Discourse Analysis involves a more detailed and iterative process. Here's an expanded explanation of each stage:

Data Collection: Begin by collecting a diverse range of texts or communication sources relevant to your research question. These sources can include written documents, spoken conversations, online forums, news articles, interviews, or any form of communication that provides insights into the topic of interest.

Transcription and Documentation: If your data includes spoken language, transcribe the spoken words into written form. Pay attention to not only the words spoken but also non-verbal elements such as pauses, tone, and gestures. For written texts, ensure you have accurate copies of the documents.

Identify Patterns: Start by identifying recurring patterns within the language used. Look for common themes, phrases, or keywords that appear frequently. Consider linguistic devices like metaphors, similes, analogies, or rhetorical strategies that convey specific meanings or emotions.

Contextual Analysis: Analyze the language within its broader context. Consider the social, cultural, and historical factors that may influence the use of certain words or phrases. Examine the power dynamics at play, including who has the authority to shape discourse and whose voices are marginalized.

Interpretation: Delve into the interpretation of language use. Try to uncover underlying meanings, ideologies, and social constructs embedded in the discourse. Question why certain words or narratives are chosen and what they reveal about the speaker's intentions or the broader cultural context.

Critical Reflexivity: Maintain a critical and reflexive stance throughout the analysis. Recognize that your perspective as a researcher can influence your interpretation. Document your own biases and assumptions and consider how they might shape your analysis.

Comparative Analysis: In some cases, it's valuable to compare different discourses or sources. Analyze how different groups or individuals frame the same topic differently. This comparative approach can reveal contrasting viewpoints and power struggles.

Ethical Considerations: Ensure that your Discourse Analysis respects ethical guidelines, especially when dealing with sensitive topics or vulnerable populations. Protect the anonymity of participants if necessary and obtain informed consent if using personal interviews or communication data.

Theory Integration: Depending on your research goals, you may integrate relevant theories into your analysis. For instance, you might draw from sociolinguistics, postcolonial theory, or critical discourse analysis to inform your interpretation.

Iterative Process: Discourse Analysis is often an iterative process. You may need to revisit earlier stages as you gain deeper insights or encounter new data that challenges your initial findings. Be open to refining your analysis as your understanding evolves.

Presentation of Findings: Present your findings in a structured manner, using examples from your data to illustrate your points. Include quotes, excerpts, or visual representations of discourse patterns to support your analysis.

Validity and Reliability: Consider how to ensure the validity and reliability of your analysis. This may involve techniques such as inter-coder reliability checks if you're working in a team or using software tools for qualitative analysis.

Discourse Analysis is a nuanced and interpretive research method that requires careful attention to language and context. The goal is to uncover the underlying structures of meaning and power within communication, providing valuable insights into how language shapes our understanding of the world.

4.8.4 How Do We Record and Store Discourse Analysis Data?

Recording and storing Discourse Analysis data involves capturing textual, spoken, or visual communication sources and organizing them for analysis. Here's a guide on how to effectively record and store Discourse Analysis data:

Recording Discourse Analysis Data:

Gather Source Material: Collect relevant sources, such as written texts, transcripts of interviews, audio or video recordings, social media content, or any form of communication that is central to your research.

Transcription: If working with spoken language, transcribe audio recordings into written text. Ensure accurate representation of speech, including pauses, tone, and non-verbal cues.

Visual Documentation: If analyzing visual materials (e.g., advertisements, photographs, artwork), ensure you have high-quality images or scans. Document any visual elements, including color, composition, and symbols.

Coding and Annotation: Develop a coding system to label and categorize segments of text or instances within the data. Annotations or comments can provide context or insights during analysis.

Metadata: Create metadata records for each source, including details like the source's origin, date, location, and relevant contextual information. This metadata helps contextualize the data during analysis.

Storing Discourse Analysis Data:

Digital Storage: Save digital data on secure computers, external hard drives, or dedicated research servers. Use strong, unique passwords for access and encrypt sensitive data.

Cloud Storage: Consider using cloud storage services for data backup and accessibility from different locations. Ensure that you understand the platform's security and privacy features.

File Organization: Develop a clear and consistent file naming and folder structure to organize your data. Include relevant details like source type, date, and context in file names.

Version Control: Implement version control practices to track changes made to your data during analysis. This helps maintain the integrity of the original data.

Data Security: Protect sensitive data by limiting access to authorized researchers and following data protection regulations. Keep backups in case of data loss or corruption.

Documentation: Maintain detailed records of your analysis process, including decisions made during coding and interpretation. This documentation aids in transparency and reproducibility.

Backup: Regularly back up your data to prevent loss due to technical failures or accidental deletions. Use multiple backup methods, including offline storage.

Data Sharing: If you plan to share your data with other researchers or publish it, ensure you have the necessary permissions and follow ethical guidelines. Anonymize or de-identify data if needed.

Data Retention: Develop a data retention policy specifying how long you will retain the data after your analysis and research project conclude. Consider ethical and legal requirements.

Data Analysis Software: Use specialized software for qualitative analysis (e.g., NVivo, ATLAS.ti) to organize and code your data efficiently and conduct in-depth analysis.

Effective recording and storage of Discourse Analysis data ensure that you can revisit and validate your findings, share insights with colleagues, and maintain the credibility of your research. Proper organization and security measures are key to preserving the integrity of the data throughout the analysis process.

4.8.5 Level of Analysis of Discourse Analysis

Table 4.2 outlines a comprehensive step-by-step process for the analysis of discourse, organized into seven levels of analysis.

At Level 1, the focus is on understanding the primary intention behind studying the text, emphasizing the need to decipher the author's or speaker's intent and meaning. Initial impressions and questions are noted.

Level 2 delves into understanding the broader context surrounding the text. This includes considerations of the background of the author or speaker and the social and political environment in which the text was produced.

Moving to Level 3, the text is categorized according to the research model, with particular attention to identifying units within the text that relate to the research constructs, dimensions, or components. The process involves assigning codes to each unit based on the research model.

Level 4 involves explicating the meanings discovered within the text, considering how the dimensions and research constructs align with the text's content. Interpretive accounts are generated, and each unit is labeled according to its relevance to the research construct.

In Level 5, linguistic features within the text units are characterized. This analysis considers aspects such as argumentation, clarity, vagueness, and transparency in the text.

Level 6 focuses on refining the components, and extrapolating interpretations onto each component of the text. It involves assessing the significance of each component for the dimension in the research model and restructuring dimensions if necessary.

Finally, Level 7 encompasses the explication of general relationships found in the model. This involves concluding the significance of each dimension and its placement within the theory used for the study. The research model may be refined based on insights gained during the analysis.

Overall, this systematic process ensures a thorough and structured approach to discourse analysis, from understanding the text's intention to refining the research model based on the insights derived from the analysis.

Table 4.2. Level of Analysis of Discourse Method

Level of Analysis	**Clarifying Questions**	**Sub-Steps**
Level 1 Understand and explicate the main intention to study the text (written, spoken, or other forms of discourse) for analysis.	How can the text be understood 'normally'? Can the statements (if spoken) be reworded or rephrased to mean something else? What does the writer/speaker wish to emphasize? How do you know this? What is the intention of the speaker/writer? How do you know this? What would be an acceptable interpretation of this text? How would you know this?	1. Read the text carefully 2. Make first notes to clarify your questions 3. Write down your first impressions of the text

(*continued*)

Table 4.2. Continued

Level of Analysis	Clarifying Questions	Sub-Steps
Level 2 Understand and explicate the context surrounding this text. Is there an organization/ person(s) behind the text creation? What is their background? What social or political context surrounds the text?	What are contextual facts that might be bridged to a story in the discourse transcript? Are there background historical or other factors known that might be evoked through the text?	3. Wherever applicable—make secondary notes—concerning the context of the discourse/text.
Level 3 Categorize the text/ discourse following the research model. What are the components or dimensions of the model being used in the study?	What are the units in the text related to the research constructs/ dimensions/components? Are there text units that can be categorized in more than one component or dimension?	4. Read discourse/ text again 5. Give codes to every text unit based on the research model 6. Compile sets of interpretive accounts per component, dimension, or construct (e.g., in tables)
Level 4 Explicate the meanings discovered in the text/ discourse	How can the dimension and the research construct be understood as a whole? How could the text accounts be estimated following the research intention, model, etc.?	7. Read the interpretive accounts that you have come up with. 8. Label every unit of analysis in line with the research intention, depicting the research construct clearly 9. Extract a core summary

Table 4.2. Continued

Level of Analysis	Clarifying Questions	Sub-Steps
Level 5 Characterize the linguistic features of the text units	What can be said about the linguistic presentation in the units? Are there convincing arguments? Clearness? Vagueness? Transparency or other linguistic features?	10. Read the interpretive accounts per set of text/discourse 11. Identify the linguistic presentation of every unit
Level 6 Refine the components—Extrapolate the interpretations on to every component of a text	What can be said about the relevancy of every component for the dimension in the research model? Are there dimensions that can be restructured after a revision?	12. Identify the noteworthiness of every text unit for the particular component analyzed 13. Identify the significance of every component for the dimension 14. Refine the structures of the dimensions (unifying some components, eliminating others, etc.)
Level 7 Explicate the general relationships found in the model	What can be concluded about significance of every dimension about their placement in the theory used for the study?	15. Refine the research model according to the insights discovered during your analysis

Adapted from Bondarouk and Ruël (2004: 14 in Sankar, 2022)

4.8.6 Applicability of Discourse Analysis

Discourse Analysis is a versatile and valuable research method applicable in a wide range of scenarios where language is a fundamental component of constructing social realities, identities, and power dynamics. Its utility extends across various academic disciplines and research contexts. Here are some specific domains and situations where Discourse Analysis finds significant applicability:

Linguistics: Discourse Analysis is a foundational approach in linguistics for understanding how language is used in various contexts, including conversations, written texts, and digital communication. Linguists employ it to explore language structures, syntax, semantics, and pragmatics.

Media Studies: In media studies, Discourse Analysis is instrumental in dissecting news articles, television programs, advertisements, and social media discourse. It helps uncover how media constructs narratives, frames issues, and influences public perception.

Sociology: Sociologists use Discourse Analysis to investigate social norms, cultural practices, and power relations within societies. It aids in understanding how language reflects and reinforces social structures.

Political Science: In political science, Discourse Analysis is employed to scrutinize political speeches, policy documents, and media coverage. Researchers can reveal how political actors frame issues, create political identities and shape public opinion through language.

Psychology: Discourse Analysis is valuable in psychology to study how individuals construct their narratives, communicate emotions, and engage in therapeutic dialogs. It aids in understanding the role of language in mental health and well-being.

Education: Educators use Discourse Analysis to investigate classroom interactions, teacher-student discourse, and the impact of language on learning outcomes. It informs pedagogical approaches and curriculum development.

Anthropology: Anthropologists employ Discourse Analysis to explore cultural practices, rituals, and ethnographic data. It helps in understanding how communities construct meaning through language.

Gender Studies: Discourse Analysis is vital in gender studies to examine how language reinforces or challenges gender stereotypes, identities, and inequalities. It reveals the role of discourse in shaping gender norms.

Environmental Studies: In environmental studies, researchers use Discourse Analysis to investigate how environmental issues are framed in policy debates, media coverage, and public discourse. It sheds light on the sociopolitical aspects of environmental problems.

Healthcare: In healthcare research, Discourse Analysis is applied to examine doctor-patient communication, medical records, and health-related narratives. It aids in improving healthcare communication and patient outcomes.

Business and Marketing: Marketers and business researchers utilize Discourse Analysis to analyze consumer reviews, advertising campaigns, and corporate communication. It helps in understanding consumer perceptions and brand messaging.

Cultural Studies: Cultural scholars use Discourse Analysis to explore how cultural artifacts, such as films, literature, and art, contribute to the construction of cultural identities and representations.

Conflict Studies: Researchers in conflict studies apply Discourse Analysis to examine narratives of conflict, peacebuilding, and reconciliation. It uncovers how language influences perceptions of conflict dynamics.

Digital Communication: Given the prevalence of online communication, Discourse Analysis is increasingly applied to analyze social media discourse, online communities, and digital narratives.

Historical Research: Historians employ Discourse Analysis to study historical texts, speeches, and archival materials, revealing how language shapes historical narratives.

In essence, Discourse Analysis is a versatile and interdisciplinary research method that provides a deeper understanding of how language functions in shaping our social, cultural, and political landscapes. Its applicability extends to any context where language is a significant factor in shaping human behavior and perception. Researchers across diverse fields harness its power to unveil the intricate connections between language, power, and social constructs.

4.8.7 Merits of Discourse Analysis

Discourse Analysis is a versatile and insightful research method that offers several merits for researchers across various disciplines and research contexts. Its unique approach to examining language and communication provides a deeper understanding of social phenomena and human interaction. Here are some key merits of Discourse Analysis:

Rich Contextual Insights: Discourse Analysis allows researchers to explore language within its social, cultural, and historical context. This contextual richness provides a nuanced understanding of how language shapes and is shaped by broader societal dynamics.

Uncovering Power Dynamics: It is particularly effective in revealing power structures and hierarchies within discourse. Researchers can identify who has the authority to speak, who is marginalized, and how power is negotiated through language.

Revealing Social Constructs: Discourse Analysis uncovers how language constructs and reinforces social realities, identities, and norms. It helps researchers examine how categories like gender, race, and class are linguistically constructed and maintained.

Exploring Subjectivity: This method acknowledges and embraces subjectivity. It recognizes that language is a product of individual perspectives and experiences, making it valuable for studying personal narratives and subjective accounts.

Holistic Understanding: Discourse Analysis offers a holistic view of communication, going beyond individual words to consider larger linguistic units, discourse patterns, and rhetorical strategies. This holistic approach is especially useful for understanding complex issues.

Adaptability to Various Data Sources: Researchers can apply Discourse Analysis to a wide range of data sources, including written texts, spoken language, interviews, media content, and visual communication. This adaptability makes it a versatile method.

Interdisciplinary Application: Its interdisciplinary nature allows researchers from diverse fields to employ this method effectively. It finds application in linguistics, media studies, sociology, psychology, anthropology, political science, and more.

In-Depth Exploration: Discourse Analysis enables researchers to delve deeply into language use, identifying recurring themes, metaphors, linguistic devices, and rhetorical strategies. This level of detail enhances the understanding of communicative phenomena.

Critical Examination of Discourses: Researchers can critically analyze dominant discourses and challenge hegemonic narratives. This is valuable for uncovering hidden biases, stereotypes, and inequalities perpetuated through language.

Evaluating Change Over Time: Discourse Analysis can be employed to track changes in language use and discourse patterns over time. This longitudinal perspective helps researchers understand evolving social norms and cultural shifts.

Policy and Practice Implications: Findings from Discourse Analysis often have practical implications for policy-making, media ethics, education, and other fields. Researchers can inform decisions and interventions based on their insights.

Researcher Reflexivity: Researchers engaging in Discourse Analysis are encouraged to reflect on their subjectivity and potential biases. This reflexivity enhances the rigor and transparency of the research process.

Engagement with Marginalized Voices: This method can amplify the voices of marginalized or silenced groups by bringing their discourse into focus. It provides a platform for underrepresented perspectives.

Enhanced Communication Skills: Conducting Discourse Analysis hones researchers' skills in critical thinking, language analysis, and interpretation, which are valuable across academic and professional settings.

Contributing to Theory Development: Discourse Analysis often contributes to the development of new theoretical frameworks and concepts, advancing knowledge within various fields.

In summary, Discourse Analysis offers researchers a multifaceted approach to studying language and communication. Its merits lie in its ability to unveil complex social constructs, power dynamics, and the intricate relationship between language and society. Researchers find value in its adaptability, interdisciplinary nature, and potential to drive meaningful change through a critical examination of discourse.

4.8.8 Demerits of Discourse Analysis

While Discourse Analysis is a valuable research method, it is not without its limitations and challenges. Researchers should be aware of these demerits to conduct robust and insightful analyzes. Here are some key demerits of Discourse Analysis:

Subjectivity and Bias: Just as it embraces subjectivity as a merit, Discourse Analysis can be criticized for subjectivity and potential researcher bias. Interpretations of discourse can vary among researchers, which might lead to different conclusions.

Time-Consuming: Conducting Discourse Analysis can be time-consuming, especially when dealing with extensive or complex datasets. The process of transcribing, coding, and analyzing language requires a substantial investment of time and resources.

Resource-Intensive: Beyond time, Discourse Analysis can also be resource-intensive, particularly when researchers need specialized software for data management and analysis. Access to transcribers or translators can also add to the cost.

Complex Analysis: The analysis of discourse can be intricate, involving coding and categorizing linguistic elements, identifying metaphors, and interpreting context. This complexity can be daunting for novice researchers.

Interpretive Challenges: Interpreting the meaning of language and discourse can be challenging, as language is often layered with multiple meanings and interpretations. Researchers may struggle to determine which interpretations are most valid.

Lack of Generalizability: Findings from Discourse Analysis might not always be easily generalizable to larger populations. The focus on language nuances and context-specific meanings can limit the broader applicability of results.

Resource and Expertise Dependence: Conducting Discourse Analysis effectively often requires specialized training and expertise in linguistics or related fields. This dependence on expertise can be a barrier for researchers without such training.

Access to Data: Accessing the necessary data sources, especially in qualitative studies, can be challenging. Researchers may face obstacles in obtaining relevant texts, recordings, or interviews.

Overemphasis on Language: Discourse Analysis primarily focuses on language as a mode of communication. This emphasis might overlook non-verbal elements of communication, such as body language or visual symbols, which can be equally important.

Limited Quantitative Insights: Discourse Analysis tends to generate qualitative insights. If researchers aim to quantify phenomena or establish statistical relationships, this method may not be suitable on its own.

Ethical Concerns: Ethical considerations related to consent, privacy, and anonymity can be complex in Discourse Analysis, especially when analyzing public or archived texts where individuals might not have anticipated research use.

Complexity of Data Sources: Analyzing mixed-media data sources, such as audio-visual content or multimodal texts, can introduce additional complexity into the analysis process.

Overreliance on Context: Interpretation often heavily relies on understanding the broader social and cultural context of the discourse. Misinterpretation can occur if researchers lack cultural competence or overlook important contextual factors.

Limited Control Over Data Creation: Researchers may have limited control over how discourse is generated, especially in the case of archived data or publicly available texts. This lack of control can impact the quality of data.

Lack of Standardization: Discourse Analysis lacks strict standardization compared to some quantitative research methods, which can make it challenging to compare studies or replicate findings.

Discourse Analysis is a nuanced research method that offers valuable insights into language and communication. However, researchers should be mindful of its limitations, including subjectivity, resource demands, and interpretive challenges, when deciding to employ this method in their research.

In summary, Discourse Analysis is a powerful method for unveiling the ways language constructs meaning, identities, and social dynamics. It offers insights into how individuals and groups use language to convey their perspectives, beliefs, and social interactions. While Discourse Analysis deepens understanding and uncovers power dynamics, researchers must navigate challenges related to subjectivity, complexity, and time demands while harnessing the benefits of this method's insights into language and society.

Here's a concise table that summarizes both the merits and demerits of discourse analysis (Table 4.3).

This table succinctly captures the strengths and limitations of Discourse Analysis, providing a clear overview for researchers considering this method for their studies. It highlights the method's versatility and depth of insight while acknowledging the challenges of subjectivity, resource demands, and interpretive complexity.

Table 4.3. Merits and Demerits of Discourse Analysis

Aspect	Merits	Demerits
Rich Contextual Insights	Provides nuanced understanding of language within social, cultural, and historical contexts.	Can be time-consuming and resource-intensive.
Uncovering Power Dynamics	Reveals power structures and hierarchies in discourse.	Subjectivity can lead to bias and different interpretations.
Revealing Social Constructs	Shows how language constructs social realities and identities.	Findings may lack generalizability and are heavily context-dependent.
Exploring Subjectivity	Embraces individual perspectives, useful for personal narratives.	Complex analysis with interpretive challenges, especially for novices.
Holistic Understanding	Considers larger linguistic units and patterns, providing a comprehensive view.	Overemphasis on language may overlook non-verbal communication elements.
Adaptability to Various Data Sources	Versatile method applicable to various data types like texts, interviews, and media.	Requires specialized training and access to diverse data sources can be difficult.
Interdisciplinary Application	Useful across multiple fields including sociology, psychology, and political science.	Lack of standardization complicates comparison and replication of studies.
In-Depth Exploration	Allows detailed analysis of language and recurring themes.	Often resource-intensive, needing specialized software and expertise.
Critical Examination of Discourses	Critically analyzes and challenges dominant narratives.	Ethical concerns, particularly around consent and privacy, can be complex.
Evaluating Change Over Time	Tracks changes in language and social norms over time.	Dependence on the broader context can lead to misinterpretation if lacking cultural knowledge.

(*continued*)

Table 4.3. Continued

Aspect	Merits	Demerits
Policy and Practice Implications	Offers practical insights for policy-making and interventions.	Limited control over data creation in public or archived texts, impacting data quality.
Researcher Reflexivity	Encourages reflection on biases, enhancing research transparency.	Can be seen as subjective, requiring careful management of researcher bias.
Engagement with Marginalized Voices	Highlights underrepresented perspectives in discourse.	Complex analysis of mixed-media sources can introduce additional complexity.
Enhanced Communication Skills	Improves critical thinking and language analysis skills.	Can be difficult to generalize findings to larger populations.
Contributing to Theory Development	Aids in developing new theoretical frameworks and advancing knowledge.	Requires significant time and resources, making it a demanding method to employ.

Source: Author

4.9 Data Collection from Secondary Sources

Data collection from secondary sources involves gathering information that has been previously collected, documented, and published by other researchers or organizations. Secondary sources provide valuable data and insights that can be analyzed and interpreted for new research purposes. This section explores the intricacies of data collection from secondary sources, covering its definition, types, the process of conducting secondary source analysis, its applicability, as well as its merits and demerits.

4.9.1 Definition of Data Collection from Secondary Sources

Data collection from secondary sources involves sourcing information from existing records, publications, databases, reports, and other sources that were created for purposes other than the current research.

4.9.2 Types of Secondary Sources

Secondary sources can take various forms, each contributing to different research objectives:

Published Literature: Academic journals, books, reports, and articles that summarize research findings, theories, and analyzes.

Official Records: Government reports, census data, official statistics, and administrative records that provide factual information.

Media Sources: News articles, broadcasts, and online content that reflect public opinions, events, and trends.

Historical Records: Archival materials, diaries, letters, and documents that offer insights into historical events and perspectives.

4.9.3 Conducting Secondary Source Analysis

The process of conducting secondary source analysis is a methodical approach to extracting knowledge and insights from existing literature, documents, or data created by others. This process is crucial for researchers seeking to build upon prior work, validate findings, or gain a deeper understanding of a particular subject. Here are the key stages involved in conducting secondary source analysis:

Source Identification:

Define Research Objectives: Clearly outline the research goals and questions that the secondary sources will address.

Source Selection: Identify relevant secondary sources such as scholarly articles, books, reports, databases, or archival materials. Consider the credibility and reputation of the sources.

Data Collection:

Access and Gather: Secure access to the selected secondary sources, either through physical copies, digital databases, libraries, or online repositories.

Proper Documentation: Maintain a comprehensive record of the sources, including publication details, publication date, authors, and source URLs, if applicable.

Evaluation:

Credibility Assessment: Evaluate the credibility and trustworthiness of each source. Consider factors such as the reputation of the author, the peer-review process (for academic papers), and the publication source.

Relevance Assessment: Assess the relevance of each source to your research objectives. Ensure that the content aligns with the specific aspects you intend to investigate.

Data Extraction:

Data Identification: Identify and extract relevant data, information, or key findings from the selected sources. This could include statistical data, quotes, arguments, or empirical results.

Organized Storage: Systematically organize the extracted data, using digital tools or note-taking techniques for easy reference during analysis.

Synthesis and Interpretation:

Data Integration: Synthesize the extracted data from multiple sources into a cohesive narrative. Identify common themes, patterns, or contradictions across the sources.

Contextual Analysis: Interpret the findings within the context of your research objectives and broader literature. Explore how the data contributes to your understanding of the research topic.

Comparative Analysis: If applicable, compare and contrast the findings or perspectives from different sources to identify areas of agreement or divergence.

Critical Reflection:

Bias Consideration: Reflect on any potential biases present in the secondary sources, including author biases, publication biases, or data collection biases.

Limitation Acknowledgment: Acknowledge the limitations of the secondary sources, such as outdated information or gaps in the existing literature.

Ethical Considerations:

Proper Attribution: Ensure that you properly attribute ideas, quotes, or data to their sources through accurate citations and references.

Avoid Plagiarism: Adhere to ethical guidelines by avoiding plagiarism and providing due credit to the authors of the secondary sources.

Report and Synthesis:

Documentation: Document your findings, interpretations, and data extractions in a structured manner, creating a reference point for your research report or paper.

Integration: Integrate the insights gained from secondary source analysis with your primary research, where applicable.

Secondary source analysis is a valuable research method that allows researchers to leverage the existing body of knowledge and build upon it. By following these systematic steps, researchers can ensure that their analysis is rigorous, and well-informed, and contributes meaningfully to the advancement of knowledge in their field of study.

4.9.4 How Do We Record and Store Secondary Source Data?

Recording and storing data from secondary sources involves gathering information from existing texts, documents, or databases and organizing it for research purposes. Here's a guide on how to effectively record and store secondary source data:

Recording Secondary Sources Data:

Source Identification: Begin by identifying and selecting relevant secondary sources that align with your research objectives. These sources can include books, articles, reports, websites, and more.

Data Collection: Collect the secondary source materials by obtaining copies of books, articles, or digital files. Ensure proper citation and referencing to maintain the source's credibility.

Summarization: Create concise summaries or abstracts of the secondary sources. Highlight key points, arguments, and findings that are relevant to your research.

Quotations: If you plan to use direct quotations from secondary sources, record these accurately, including page numbers for reference.

Digital Formats: If dealing with digital sources, you can copy and paste relevant text into a digital document or note-taking software. Ensure proper citations for each excerpt.

Storing Secondary Sources Data:

Digital Storage: Save digital secondary source materials on secure computers, external hard drives, or cloud storage services. Use strong, unique passwords for access and encrypt sensitive data if necessary.

File Organization: Develop a clear and consistent file naming and folder structure to organize your secondary source data. Include details like author names, publication dates, and source titles in file names.

Reference Management Software: Consider using reference management software (e.g., Zotero, EndNote, Mendeley) to store and organize citations and PDFs of articles or documents. These tools can help streamline the referencing process.

Version Control: If you make annotations or notes on secondary source materials, ensure you have a version control system in place to track changes and revisions.

Metadata: Create metadata records for each secondary source, including details like the author, publication date, source title, and a summary. This metadata helps you quickly locate and cite sources.

Backups: Regularly back up your secondary source data to prevent loss due to technical failures or accidental deletions. Use multiple backup methods, including offline storage.

Data Sharing: If you plan to share your secondary source data with other researchers, consider creating a dataset with proper documentation and permissions. Ensure compliance with copyright and licensing agreements.

Data Retention: Develop a data retention policy specifying how long you will retain the secondary source data after your research project concludes. Consider ethical and legal requirements.

Data Analysis Software: If you are using secondary sources for data analysis, use appropriate software tools (e.g., statistical software, qualitative analysis software) to work with the data effectively.

Effective recording and storage of secondary source data ensure that you can easily access, reference, and cite the materials during your research. Proper organization and security measures help you maintain data integrity and comply with ethical and legal guidelines.

4.9.5 Applicability of Data Collection from Secondary Sources

Data collection from secondary sources is a valuable research approach applicable in a wide range of scenarios across various disciplines. Researchers often turn to existing data and literature when conducting studies. Understanding the applicability of this method is essential for designing robust research projects. Here are key scenarios where data collection from secondary sources finds relevance:

Historical Research: Data collection from secondary sources is instrumental in historical studies. Researchers rely on archived documents, records, newspapers, and historical literature to reconstruct events, social dynamics, and cultural phenomena from the past. This approach offers valuable insights into bygone eras.

Literature Reviews: Secondary sources are fundamental in literature reviews. Researchers use them to synthesize existing knowledge, identify research gaps, and build theoretical frameworks. This method helps in comprehensively understanding the state of the art in a particular field.

Longitudinal Studies: For research projects spanning extended periods, data collection from secondary sources allows for the analysis of trends and changes over time. Datasets, reports, and documents from different time points can be compared to assess developments.

Comparative Analyzes: Researchers often use secondary data to conduct comparative analyzes. This involves comparing data from different geographic locations, populations, or periods to identify similarities, differences, and potential causal relationships.

Policy Analysis: Government reports, legislative documents, and official statistics serve as valuable secondary sources for policy analysis. Researchers assess the impact of policies, track changes in regulations, and evaluate the effectiveness of interventions.

Economic Research: Economists frequently use secondary data sources, including financial reports, stock market data, and government economic indicators, to analyze economic trends, market behavior, and economic performance.

Social Sciences: In sociology, anthropology, and related fields, researchers draw on secondary data to explore social phenomena, cultural practices, and demographic trends. This approach is useful for cross-cultural studies and examining societal shifts.

Environmental Studies: Environmental researchers access secondary data to study ecological changes, climate trends, and biodiversity. Archived environmental records, satellite imagery, and climate data contribute to comprehensive analyzes.

Healthcare and Epidemiology: Secondary data, such as medical records, disease registries, and healthcare databases, are essential for epidemiological studies, health assessments, and tracking disease outbreaks.

Market Research: Businesses and marketers rely on secondary sources to gather information about consumer behavior, industry trends, and competitor analysis. Market reports, industry studies, and customer reviews are valuable resources.

Education Research: Researchers in education utilize secondary data sources, including standardized test scores, school records, and educational surveys, to assess educational outcomes, performance trends, and curriculum effectiveness.

Cross-disciplinary Studies: Secondary data can facilitate cross-disciplinary research by integrating information from multiple sources and fields to address complex issues that require diverse expertise.

Resource and Cost Efficiency: Data collection from secondary sources can be more resource and cost-efficient compared to primary data collection, making it an attractive option for research projects with budget constraints.

Inaccessible or Sensitive Data: In situations where primary data collection is not feasible due to geographic remoteness, ethical concerns, or inaccessibility to specific populations, secondary data offers a viable alternative.

Understanding when and how to leverage data from secondary sources is a skill that enhances the robustness and efficiency of research endeavors. Researchers must critically assess the quality, reliability, and relevance of secondary data to ensure its suitability for addressing research questions and advancing knowledge in their respective fields.

4.9.6 Merits of Data Collection from Secondary Sources

Data collection from secondary sources, which involves utilizing existing data and literature for research purposes, offers numerous advantages that make it a valuable research approach in various disciplines. Understanding these merits is essential for researchers to effectively harness the potential of secondary data. Here are the key advantages:

Cost-Efficiency: One of the most significant advantages of using secondary data is its cost-effectiveness. Unlike primary data collection, which often requires substantial resources, secondary data is typically readily available or can be obtained at a lower cost. This cost efficiency is particularly beneficial for researchers with limited budgets.

Time Savings: Collecting primary data can be time-consuming, involving tasks such as survey administration, data entry, and participant recruitment. Secondary data is readily accessible, allowing researchers to save time and focus on analysis and interpretation.

Large-Scale Data: Secondary sources often provide access to large-scale datasets that may be impractical or impossible to collect through primary means. These extensive datasets can offer comprehensive insights and support in-depth analyzes.

Longitudinal Analyzes: Secondary data can encompass historical records and longitudinal datasets, enabling researchers to conduct analyzes over extended periods. This longitudinal perspective is valuable for studying trends, changes, and developments over time.

Comparative Studies: Researchers can use secondary data to conduct comparative studies across different geographical regions, populations, or periods. This approach allows for the exploration of variations, similarities, and differences, contributing to a richer understanding of research topics.

Validation and Replication: Secondary data can be used to validate or replicate previous studies' findings. This process enhances the robustness of research by confirming or challenging existing knowledge.

Access to Inaccessible or Sensitive Data: In cases where primary data collection is not feasible due to ethical concerns, logistical challenges, or inaccessibility to certain populations, secondary data can serve as a valuable alternative.

Diverse Data Types: Secondary data encompasses a wide range of data types, including quantitative, qualitative, textual, and multimedia data. Researchers can select data sources that align with their research objectives and methods.

Interdisciplinary Research: Secondary data can facilitate interdisciplinary research by integrating data from multiple sources and fields. This interdisciplinary approach is especially beneficial for addressing complex issues that require diverse expertise.

Reduced Participant Burden: Using secondary data eliminates the need to involve human participants, reducing ethical and logistical considerations related to informed consent, privacy, and confidentiality.

Resource Availability: Many reputable organizations, institutions, and government agencies collect and maintain extensive datasets, making them available to researchers. These resources provide valuable insights into various domains.

Enhanced Reliability: Established and standardized data collection procedures used in secondary sources can enhance the reliability and credibility of research findings. These procedures often undergo rigorous quality control processes.

Ethical Considerations: Secondary data collection typically involves fewer ethical dilemmas and risks compared to primary data collection, as it does not directly involve human subjects.

Environmental Impact: Data collection from secondary sources contributes to reducing the environmental impact associated with primary data collection methods that require resources like paper, transportation, and energy.

Understanding these merits of data collection from secondary sources empowers researchers to make informed decisions about the most suitable data collection approach for their research questions. However, it's crucial to acknowledge that secondary data also has limitations, and researchers must critically evaluate its quality and relevance for their specific research objectives.

4.9.7 Demerits of Data Collection from Secondary Sources

While data collection from secondary sources offers numerous advantages, it is not without its limitations and challenges. Researchers must consider these demerits when utilizing secondary data for their studies:

Data Quality Concerns: The quality of secondary data can vary widely. Researchers may encounter issues related to accuracy, completeness, reliability, and consistency. Data collected for purposes other than the current research may contain errors, biases, or inconsistencies that can affect the validity of findings.

Limited Control: Researchers have limited control over the data collection process when using secondary sources. They cannot tailor data collection instruments or methodologies to their specific research objectives, potentially resulting in gaps or inadequacies in the data.

Availability and Accessibility: Not all data may be readily available or accessible. Some datasets may be proprietary, confidential, or restricted due to privacy or legal considerations. Researchers may encounter challenges in obtaining the desired data, limiting the scope of their research.

Relevance and Fit: Secondary data may not always align perfectly with the research questions or objectives. Researchers must carefully assess whether the available data is relevant, appropriate, and sufficient to address their specific research needs.

Data Bias: Secondary data may reflect biases or assumptions inherent in the original data collection process. These biases can stem from sampling methods, data collection instruments, or the context in which the data was collected. Researchers must be vigilant in identifying and addressing potential biases.

Data Currency: Data collected from secondary sources may become outdated or less relevant over time. Researchers must consider the currency of the data, especially when analyzing trends or conducting longitudinal studies.

Data Limitations: Secondary data may lack certain variables or information that researchers require for their analysis. Researchers may need to make assumptions or compromises in their research due to data limitations.

Heterogeneity: Datasets from secondary sources may come from diverse origins, introducing heterogeneity into the data. Variability in data sources, formats, and definitions can pose challenges in data integration and analysis.

Ethical Considerations: Ethical concerns related to the original data collection may persist when using secondary data. Researchers should be aware of any ethical issues associated with the data source and take steps to address them, such as ensuring privacy and confidentiality.

Data Ownership and Permissions: Researchers must navigate issues of data ownership and permissions when using secondary data. They should seek appropriate permissions or licenses when required, especially for copyrighted or proprietary data.

Limited Customization: Secondary data may not allow for customization of variables or data collection methods to suit specific research objectives. Researchers may need to work within the constraints of the available data.

Data Documentation: Incomplete or inadequate documentation of the secondary data source can hinder the research process. Researchers may struggle to understand data variables, definitions, and coding schemes.

Data Integration Challenges: When combining multiple secondary datasets, researchers may encounter challenges related to data integration, harmonization, and standardization. Inconsistent formats and units of measurement can complicate this process.

Loss of Context: Secondary data may lack the contextual information and nuances that researchers would have obtained through primary data collection methods, potentially limiting the depth of analysis.

Researcher Expertise: Effectively utilizing secondary data requires specific data analysis skills and domain expertise. Researchers must be proficient in data manipulation, statistical techniques, and the interpretation of existing datasets.

Despite these demerits, secondary data remains a valuable resource for researchers, offering opportunities to explore a wide range of research questions and topics. Researchers can mitigate some of these challenges through careful data selection, validation, and transparent reporting of limitations in their research findings.

In summary, data collection from secondary sources offers a valuable means of accessing existing information and insights for new research endeavors. It provides time and cost efficiencies while enabling historical, comparative, and longitudinal analyzes. While secondary sources contribute to the research process, researchers must navigate challenges related to source quality, context, and ethical considerations while harnessing the benefits of this method's efficient access to existing knowledge and data (For details, please see Islam & Siti Hajar, 2013; Islam et al., 2014, 2021a, 2021b).

CHAPTER 5

Designing Data Collection Instruments

ABSTRACT

This chapter serves as a practical guide for crafting effective tools to gather data. It encompasses a diverse array of instruments, from interview schedules and guidelines to checklists for observation, focus group discussions (FGDs), participatory rural appraisal (PRA), discourse analysis, key informant interviews (KIIs), community mapping, and secondary data sources. This chapter equips researchers with templates and samples to create instruments tailored to their specific research objectives.

Keywords: Data collection instruments, interview schedule, research guidelines, observation checklist, focus group discussions, participatory rural appraisal, discourse analysis, key informant interviews, community mapping, secondary data sources

5.1 Data Collection Instruments: Schedule, Guideline, and Checklist

Effective data collection instruments are essential for ensuring consistency, accuracy, and reliability in data collection processes. This section focuses on the design and components of data collection instruments, including schedules, guidelines, and checklists. These instruments provide structure, guidance, and a standardized approach to data collection, contributing to the overall quality of research outcomes.

5.1.1 Data Collection Instruments: Overview

Data collection instruments encompass various tools and documents that facilitate the collection of data. They guide researchers, interviewers, or fieldworkers in conducting surveys, interviews, observations, and other data collection methods. Three key types of data collection instruments are schedules, guidelines, and checklists (Islam, 2019, 2022b). The preparation of data collection instruments lacks a one-size-fits-all rule or template. Instead, the process's specifics are contingent on various factors, including the chosen data collection method, the specific data collection tool, and the overarching research objectives. This dynamic nature necessitates flexibility and adaptability in designing these instruments to best suit the unique requirements of each research endeavor.

5.1.2 Data Collection Schedules

Data collection schedules outline the sequence and timing of data collection activities. They provide a structured plan for when and how data will be collected. Schedules are particularly crucial for longitudinal studies or projects involving multiple phases. A well-designed schedule ensures that data collection is organized, systematic, and aligned with research objectives. Scheduled are used for structured and semi-structured interviews which are called interview schedules.

5.1.3 Data Collection Guidelines

Data collection guidelines offer instructions, prompts, and tips for researchers or interviewers conducting data collection. Guidelines provide a consistent approach to asking questions, interacting with participants, and handling various scenarios that may arise during the data collection process. They ensure uniformity and help minimize bias and errors. Guidelines are mostly used in FGDs, in-depth case interviews, PRA, and discourse analysis.

5.1.4 Data Collection Checklists

Data collection checklists serve as a quality assurance tool. They outline essential steps, materials, and considerations for conducting data collection methods accurately and effectively. Checklists help researchers ensure that all necessary components are in place before, during, and after data collection. Checklists are used in observation, KIIs, community mapping, and secondary sources.

5.1.5 Aspects of Data Collection

When embarking on the design of data collection instruments, it's imperative to pay close attention to several pivotal aspects:

Clarity and Simplicity: The effectiveness of your instruments hinges on their ability to communicate clearly and succinctly. Instructions, questions, and prompts should be crafted in a manner that is easily comprehensible not only to data collectors but also to the participants themselves.

Consistency: To maintain a high level of data integrity, it's essential to uphold consistency in the wording, formatting, and stylistic elements throughout the entire instrument. This uniformity mitigates confusion and ensures that data is collected in a standardized fashion.

Relevance: Every question and prompt within the instrument should directly tie back to the overarching research objectives. The content must align seamlessly with the focus of the study to yield meaningful insights.

Avoid Leading Questions: Bias has no place in effective data collection. Questions should be phrased in a neutral manner that doesn't inadvertently steer participants toward particular responses. This neutrality fosters genuine, unbiased data.

Pretesting: Before locking in your instruments, it's advisable to conduct pretests with a small sample of participants. This crucial step helps unearth any potential issues, ambiguities, or misunderstandings within the instruments, enabling refinements to be made.

Cultural Sensitivity: Recognize the cultural context in which your instruments will be employed. Ensure that they are not only sensitive but also appropriate for the target audience. Cultural nuances can significantly impact the effectiveness of data collection, so it's vital to take them into account during the design process.

5.1.6 Benefits of Well-Designed Instruments

Well-designed data collection instruments confer numerous advantages to the research process:

Accurate Data Collection: Clear and precise instruments lead to accurate data collection. Participants and data collectors can easily comprehend and respond to questions, reducing errors and ambiguities.

Consistency: Consistency in instrument design ensures uniform data collection across different settings, time points, or data collectors. This consistency enhances data reliability and comparability.

Relevance: Instruments tailored to the research objectives guarantee that collected data is directly aligned with the study's focus. This relevance enables researchers to extract meaningful insights.

Minimized Bias: Well-structured instruments, free from leading or biased questions, reduce the risk of bias in responses. Participants are more likely to provide genuine and unbiased information.

Efficiency: Clear and straightforward instruments streamline the data collection process. Data collectors can efficiently administer the instruments, saving time and resources.

Prevention of Ambiguities: Through pretesting, well-designed instruments can identify and rectify potential ambiguities or misunderstandings. This proactive approach helps prevent data collection hiccups.

Enhanced Participant Engagement: When participants find instruments easy to understand and relevant, they are more likely to engage actively in the data collection process. This engagement can lead to more comprehensive and accurate data.

Cultural Sensitivity: Instruments that consider cultural context and sensitivities enhance rapport with participants from diverse backgrounds. Cultural sensitivity fosters trust and facilitates data collection in culturally nuanced settings.

Data Quality: The cumulative effect of these benefits is an improvement in data quality. High-quality data is the foundation for robust analysis and meaningful research findings.

Cost Savings: Well-designed instruments reduce the likelihood of errors, rework, or additional data collection efforts. This can lead to cost savings in terms of both time and resources.

In summary, data collection instruments play a pivotal role in ensuring the quality, consistency, and reliability of collected data. Schedules, guidelines, and checklists provide structure and guidance, enabling researchers to systematically collect accurate and relevant data. Well-designed instruments contribute to the success of research projects by facilitating efficient and effective data collection processes.

5.2 Sample of Interview Schedules (Quantitative and Qualitative Surveys)

5.2.1 Example 1: Structured Interview Schedule for Quantitative Survey

Household Survey: Structured Questionnaire for Interview

POVERTY AND INEQUALITY IN BANGLADESH: REGIONAL DISPARITY IN PARTICULAR REFERENCE TO THE SUSTAINABLE DEVELOPMENT GOALS (SDGS)

[The main objective of this research is to explore the regional disparity of poverty and inequality in terms of income, land ownership, education, and social opportunities in Bangladesh in particular regarding the Sustainable Development Goals (SDGs). This research is funded by the Ministry of Education, People's Republic of Bangladesh, and organized by the Bangladesh Bureau of Educational Information and Statistics (BANBEIS). I will be grateful if you share your opinions regarding this topic. Please note that the information given by you will be used only for research purposes, and I will not disclose your opinions I will strictly maintain the confidentiality of your information. You have every right to withdraw yourself from the interview at any time. After finishing the interview, I will let you know what I have written to you and you will have the opportunity to add or delete any of your opinions/information that you will provide. Your cooperation would be highly appreciated]

Q. Code # .... Name of the Interviewer: ....
Location: Urban:......... Rural:............ Date of interview: ....
Cell phone number. ....

1. Demographic Information

1.1	**1.1.1 Name of the Household Head:** .... **1.1.2 Gender:** Male =**1**, Female =**2** (If the household head is not found, then write the name of the respondent: ....) **1.1.3 Relationship with the Household head:**								
1.2 Address: Village			Union						
Upazilla			District			Division			
1.3 HHs Detail Information									
Sl	HHs Members Name (From elder to younger)	Relationship with Respondent	Gender (Male = 1 Female= 2)	Marital status: (Unmarried = 1 Married = 2 Divorced = 3 Widowed = 4 N/A = 5)	Age	Education (No. of Years of schooling, can sign = 99, can'tsign=0)	Main occupation (See below code)	Secondary occupation (See below code)	Average monthly income (in Tk.)
1	2		3	4	5	6	7	8	9
1.	(HH)								
2.									

3.									
4.									
5.									
6.									
7.									
8.									
								Total average monthly income of the Family	

Main & Secondary occupation code: Day laborer =**1**, Agriculture = **2**, Fishing = **3**, Govt. Job = 4, Business = 5, Farming = 6, Rickshaw/van pulling = 7, Driver of motor bike/taxi = 8, Grocery shop = 9, Tailoring = 10, Hawker = **11**, Maid servant = **12**, Teacher/house tutor = **13**, Student = **14**, NGO job =15, Garments worker = 16, Other = 17 (Please specify here. . . . .)

1.4 Religious faith of the respondent	Islam = **1**, Hindu = **2**, Christian = **3** Buddhist = **4**, Others = **5** ------ ------ ------ ------ ------ -----

2. ECONOMIC PROFILE OF THE HHs (Occupation-Income-Expenditure-Asset-Land)		
2.1 List of assets		
Type of asset	**Amount/number**	**Approximate value** (in Tk.)
1. Cash capital		
2. Invested resources		
3. Land		
4. Ornaments		
5. Livestock (mention the type of livestock separately in the next column)		
6. House property		
7. Vehicle		
8. Trees		
9. Shop		
10. Others (specify)		
% of ownership	**Husband**........%	**Wife**...........%
2.2 Sources of Income		

Sources of earnings	Sub-group Specific name of the trade/ farming/service etc.	Specific Name of the trade/ crop/ farming/ services, etc. Yes = 1, No = 2	Average monthly income (Taka)
1	2	3	4
Agricultural Activities Farming	1. Crop cultivation		
	2. Dairy		
	3. Fishery		
	4. Beef fattening		
	5. Goat/sheep rearing		
	6. Poultry/chicken rearing		
	7. Other farming		

Non-agricultural Activities	8. Sewing garments	Service or Job	
	9. Service (Job)	Various Job holders	
	10. Day laborer	Day laborer in other business Day laborer in agricultural field	
	11. Driving	Rickshaw /van/ auto/ motor-cycle/ boat driving/ rent/ others	
	11. Business	[Mention the name of the business sector]	
	12. Small Business	Tea stall, Grocery, Selling trees and Other types of Small Business	
	13. Financial assistance from family members	[Relationship]	
	14. Remittance	Earn from a foreign country by family members or relatives	
Support	15. Govt. pension	Government pension	
	16. Zakat	Zakat and Fitra received	
	17. Support received	Old Age Allowance	
	18. Support received	Widow allowance	
	19. Support received	Freedom fighters allowance	
	20. Support received	Stipend	
	21. Support received	Relief	
	22. Support received	Cash for work/ training	
	23. Support received	Disability allowance	
	24. Support received	VGF	
	25. Support received	VGD	
Total % of financial value own	**Husband** . . . . % **Wife** . . . .%		

2.3 Expenditure	
Types of expenditure	**Total amount (Monthly Average over 12 months)**
1. Spending for own recurrent capital like raw materials for business/agricultural inputs/pesticide/etc.	
2. Food	
3. House	
4. Education	
5. Health services	
6. Purchasing clothing	
7. Electricity/gas/water	
8. Entertainment	
9. Transport	
10. Loan repayment (if any)	
11. Recreation and festivals	
12. Dowry	
13. Others	
Total	
2.4 Savings (last one year)	
1. Cash in hand	
2. Cash at Bank	
3. Cash at NGO savings	
4. Others	
Total	
2.5 Debt (last one year)	
1. Bank loan	
2. Loan from friends/relatives	
3. Loan from money lenders	
4. Loan from NGO/micro credit	
5. Debt to shopkeepers/suppliers	
6. Others (Specify)	
Total	

2.6 Access to Landholdings (agricultural and dwelling)				
2.6.1	Do you possess any land?		Yes = 1	No = 2
2.6.2	If yes, please describe the land			
	Types of land	Total amount of land (Decimal)	Land used (Decimal)	
Sl	1	2	3	
	Own			
	Lease out			
	Lease in			
	Khash			
	Sharing			
	Others (Specify)			
	Total			
	Used of land code: Homestead = **1**, Cultivable = **2**, Pond = **3**, Uncultivable land = **4**, Others = **5**			

3. FOOD SECURITY			
3.1	Do you have three meals a day regularly?	Yes = **1**	No = **2**
3.2	Food Security Profile (status by month) Meal Frequency Code: **1** = Once, **2** = Twice, **3** = Thrice		

Months	**Meal Frequency**	**Months**	**Meal Frequency**
1. Baishak (April–May)		7. Kartik (Oct–Nov)	
2. Jiashtha (May–June)		8. Augrahayan (Nov–Dec)	
3. Ashar (June–July)		9. Poush (Dec–Jan)	
4. Sravan (July–August)		10. Magh (Jan–Feb)	
5. Bhadra (August–Sept)		11. Falgun (Feb–Mar)	
6. Ashwin (Sept–Oct)		12. Chaitra (Mar–April)	

4. HOUSING, WATER SOURCES, LATRINE, AND SANITATION PRACTICES					
4.1 What is the ownership status of your house?					
Owned = 1	Rented = 2	Mortgaged = 3	Living free with neighbor/ relative = 4		Others (Please specify) = 5. . . . .
4.2 What is the type of your housing structure?					
Cemented Building (Pacca) = 1	Semi-pucca building with Tin roof = 2	Thatched house = 3	Tin/Wood/ Bamboo = 4	Mud/ straw = 5	Others (Please specify) = 6
4.3 Who helped to make your housing?					
Inherited = 1	Own earning = 2	Donated by government organization = 3	Donated by NGOs = 4	Personal/ relative donation = 5	Other (please specify) = 6
4.4 What is the condition of the household housing? (Please probe)					
Excellent = 1	Good = 2	Moderate = 3	Fair = 4	Bad = 5	Very bad = 6
4.5 What type of toilet facility does your household usually use? (Probe if possible)					
None (open defecation) = 1			Sanitary Latrine = 4		
Communal, enclosed pit = 2			Slab with Ring = 5		
Private, enclosed pit = 3			Hanging latrine = 6		
4.6 What is the main source of water for your household uses? Who owns the source? How far do you travel to fetch water?					
Types	Source of water	Ownership	Distance travel (Yard)	Sufficient to meet your requirements (Yes = 1 / No = 2)	Safety issue for females to fetch water—(Yes = 1/ No = 2)
1	2	3	5	6	7
1. Drinking					
2. Cooking					
3. Bathing					
4. Washing					
5. Toilet					

Source of water code: Deep Tube-well = 1, Shallow Tube-well = 2, Supply = 3, Harvested Rain water = 4, Pond = 5, Cannel = 6, River = 7
Ownership Code: Fully Own = 1, Shared = 2, Relatives = 3, Community = 4, Government = 5, Neighbor = 6

1	Households having hand washing devices	HH having HWD [ ]	HH without HWD [ ]
2	Households use safe water for all purposes (bathing, cooking, and drinking) (put tick)	All the people [ ] Most of the people [ ] Half of the people [ ] Some of the people [ ]	
3	Households wash hands with soap after defecation and before having meal	All the people [ ] Most of the people [ ] Half of the people [ ] Some of the people [ ]	

5. ACCESS TO COMMUNITY/SOCIAL OPPORTUNITIES						
Types	Level of access					
	Very high	High	Average	Lower	Very lower	No access at all
1. Union Parishad						
2. River and river-related resources such as fishing, sand, communication, etc.						
3. Open spaces						
4. Community Center						
5. Public toilet and tube well						
6. Mosque/temple						
7. Market						
8. Club						
9. Embankment/canal						
10. Mash and fishing/pond						
11. Associations such as youth associations, voluntary efforts, women's association						
12. Trees (government)						
13. Park/garden						
14. Use of local institutions such as schools for individual or community interest if needs						
15. Hat-Bazar						

16. Playground					
17. Social/cultural activities					
18. Kashland					
19. Water					
20. Common religious property/premises					
21. Community clinic					
22. Dakbanagla/Rest house					
23. Graveyard/Shashan					
24. Hill					
25. Forest road side					
26. Other (identify)					

6. PUBLIC OPINIONS ON ECONOMIC, SOCIAL, PSYCHOSOCIAL, CULTURAL, AND POLITICAL PERSPECTIVES ON POVERTY AND INEQUALITY

Type of	**Select and circle the response category that best represents your reaction to each statement**						
	Strongly Agree	**Agree**	**Somewhat agree**	**Somewhat disagree**	**Disagree**	**Strongly disagree**	**No comment**
I feel that I get a proper opportunity in the society to increase my income	1	2	3	4	5	6	7
I think that there is no big gender gap in Bangladesh	1	2	3	4	5	6	7
I observe that there are no differences in terms of gaining opportunities among different age groups	1	2	3	4	5	6	7
I think the disabled people are getting equal opportunities with the normal citizen	1	2	3	4	5	6	7
I think there is no difference in terms of getting access to rights and opportunities among race, class, and ethnicity	1	2	3	4	5	6	7
I think there is no big difference between urban and rural in terms of income, education, and other opportunities	1	2	3	4	5	6	7
I think all religious people are equally treated and have equal opportunity in this country	1	2	3	4	5	6	7

I think the rich class is getting more opportunities in the society than the poor class	1	2	3	4	5	6	7
I think children in rural areas are more vulnerable than the urban							
I think the social protection system is equal between rural and urban	1	2	3	4	5	6	7
I think the poor and the vulnerable have equal rights to economic resources as well as acccss to basic services	1	2	3	4	5	6	7
I think the poor and the vulnerable have equal ownership and control over land and other forms of property	1	2	3	4	5	6	7
I think the poor and the vulnerable have equal rights to inheritance resources and capital such as land	1	2	3	4	5	6	7
I think the poor and the vulnerable have equal rights to natural resources	1	2	3	4	5	6	7
I think the poor and the vulnerable have equal rights to appropriate new technology	1	2	3	4	5	6	7
I think the poor and the vulnerable have equal rights to financial services, including microfinance	1	2	3	4	5	6	7

I think the government is giving equal priority between urban and rural to build the resilience of the poor and those who are vulnerable	1	2	3	4	5	6	7
I think the government is giving equal priority between urban and rural poor people to reduce vulnerability to climate-related extreme events and other economic, social, and environmental shocks and disasters	1	2	3	4	5	6	7
I think the government is ensuring equal significance of the mobilization of resources from a variety of sources	1	2	3	4	5	6	7

Thank you for your valuable responses

Source: Islam (2022b)

5.2.2 Example 2: Semi-structured Interview Schedule for Qualitative Survey

Research Title: Exploring Factors Influencing Job Satisfaction among Healthcare Professionals

Introduction:

Introduce yourself and explain the purpose of the interview.
Obtain informed consent from the participant.
Assure confidentiality and anonymity of responses.

Section 1: Demographic Information

- Age:
- Gender:
- Occupation:
- Years of experience in the healthcare industry:
- Marital status:

Section 2: Family and Social Support

- Could you briefly describe your family structure and dynamics? (Open-ended)
- How does your family support or impact your job satisfaction? (Open-ended)
- Are there any family-related factors that influence your job satisfaction positively or negatively? (Open-ended)

Section 3: Financial Aspects

- Could you share your thoughts on how your income level affects your job satisfaction? (Open-ended)
- Do you feel your income is commensurate with your qualifications and job responsibilities? (Closed: Yes/No, Open-ended follow-up)
- How do financial considerations play a role in your overall job satisfaction? (Open-ended)

Section 4: Assets and Work-Related Investments

- Do you perceive your workplace to offer adequate benefits or non-monetary assets (e.g., healthcare benefits, flexible hours, professional development opportunities)? (Closed: Yes/No, Open-ended follow-up)
- Could you provide an example of how a specific work-related asset has contributed to your job satisfaction? (Open-ended)

Section 5: Work-Life Balance and Financial Pressure

- How does your financial situation impact your ability to achieve work-life balance? (Open-ended)
- Are there instances where financial concerns have affected your job satisfaction or performance? (Open-ended)

Section 6: Suggestions and Reflections

In your opinion, what measures could be implemented to address financial-related factors that influence job satisfaction among healthcare professionals? (Open-ended)

Conclusion:

Thank the participant for their time and insights.

- ○ Address any questions or concerns the participant might have.
- ○ Reiterate the confidentiality of responses.

[This extended interview schedule includes questions related to family dynamics, income, assets, and their connections to job satisfaction. Researchers can adapt these questions to align with their specific research objectives and the characteristics of the study population. By including questions about family, income, and assets, this interview schedule aims to explore the multifaceted influences on job satisfaction among healthcare professionals.]

Source: Author

5.3 Sample Checklist for Observation

An observation checklist serves as a tool to systematically record specific behaviors, actions, and events during an observation session. It ensures that important aspects are noted and provides a structured framework for data collection. Below is a sample checklist for conducting observations in a classroom setting:

Observation Checklist

Research Title: Exploring Classroom Engagement Strategies
Observation Date: [Date of Observation]
Observer: [Name of Observer]
Objective: To observe and document classroom engagement strategies employed by the teacher during a middle school science lesson.
Classroom Setting: Middle School Classroom, Grade 8
Teacher: [Name of Teacher]
Duration of Observation: [Duration of Observation]

Checklist:

Classroom Environment:

- □ Seating arrangement promotes student interaction.
- □ Classroom displays include relevant science-related materials.
- □ The classroom is organized and free from distractions.

Teacher's Interaction:

- □ Greets students as they enter the classroom.
- □ Maintains eye contact and a positive demeanor while addressing students.
- □ Encourages students to ask questions.

Lesson Content:

- □ Presents clear learning objectives at the beginning of the lesson.
- □ Explains the relevance of the lesson to students' lives.
- □ Relates the lesson content to real-world examples.

Engagement Strategies:

- ☐ Ask open-ended questions that promote critical thinking.
- ☐ Uses interactive activities such as group discussions and peer teaching.
- ☐ Incorporates hands-on experiments or demonstrations.
- ☐ Provides opportunities for student input and contributions.

Student Engagement:

- ☐ Students actively participate in discussions.
- ☐ Students respond to questions with thoughtful answers.
- ☐ Students collaborate and share ideas during group activities.

Teacher's Response:

- ☐ Acknowledges and validates students' contributions.
- ☐ Provides constructive feedback and encouragement.
- ☐ Adjusts the pace of the lesson based on student responses.

Classroom Management:

- ☐ Manages disruptions promptly and respectfully.
- ☐ Uses positive reinforcement to encourage desired behavior.
- ☐ Ensures all students have an opportunity to participate.

Use of Technology:

- ☐ Incorporates multimedia resources to enhance learning.
- ☐ Uses technology to display visuals and support explanations.

Conclusion:

- ☐ Summarizes key points and learning objectives at the end of the lesson.
- ☐ Provides students with a clear understanding of homework or assignments.

Overall Assessment:

- ☐ Engaged and interactive classroom environment.
- ☐ Effective use of engagement strategies to maintain student interest.
- ☐ Positive teacher-student interactions.

[This observation checklist offers a structured framework for capturing various aspects of classroom engagement strategies and teacher-student interactions. Researchers and observers can customize the checklist to suit the research objectives and specific classroom context. By using this checklist, observers can systematically document behaviors and events that contribute to the overall understanding of engagement strategies employed in the observed classroom setting.]

Source: Author

5.4 Sample Guideline for Focus Group Discussions (FGDs)

Focus Group Discussions (FGDs) provide a platform for group interactions and in-depth discussions on specific topics. A well-structured FGD guideline ensures that participants explore the research topic thoroughly and share their perspectives openly. Below is a sample guideline for conducting a focus group discussion on the theme of "Youth Perspectives on Environmental Sustainability."

Focus Group Discussion (FGD) Guideline

Research Title: Exploring Youth Perspectives on Environmental Sustainability

Introduction:

Welcome participants and introduce yourself as the facilitator.

Explain the purpose of the focus group and the research objectives.

Emphasize the importance of open and respectful discussion.

Icebreaker: Begin with a brief icebreaker activity to help participants feel comfortable and get to know each other.

Objective: To gather insights and perspectives from young adults on their attitudes, behaviors, and concerns related to environmental sustainability.

Focus Group Setting: Youth Community Center, Meeting Room

Date and Time: [Date and Time of FGD]

Duration: [Duration of FGD]

Facilitator: [Name of Facilitator]

Recording: Explain that the FGD will be audio-recorded for accurate documentation.

Guideline: Start by discussing the concept of environmental sustainability and its importance.

- Ask participants to share their general thoughts and attitudes toward environmental sustainability.
- Explore individual practices: What are some environmentally friendly behaviors participants engage in regularly?
- Discuss challenges: What obstacles or barriers do participants face in adopting sustainable behaviors?
- Delve into motivations: What factors drive participants to make sustainable choices? Are there personal values or external influences?
- Explore concerns: What environmental issues do participants find most concerning? Why?
- Discuss community involvement: Are there any community initiatives or projects related to sustainability that participants are engaged in or aware of?
- Introduce hypothetical scenarios: Present scenarios related to environmental choices (e.g., plastic use, energy consumption) and ask participants how they would respond.
- Explore communication channels: How do participants stay informed about environmental issues? Do they share information with others?
- Encourage group interactions: Invite participants to respond to each other's comments, share similar or contrasting experiences, and ask follow-up questions.
- Ensure inclusivity: Make sure all participants have an opportunity to contribute and that no one dominates the discussion.

- Summarize key points: Recap the main insights shared by participants before concluding the FGD.
- Conclusion:
  - ✓ Thank the participants for their valuable contributions and time.
  - ✓ Explain the next steps of the research process.
  - ✓ Assure participants that their responses will be treated confidentially.

[This focus group discussion guideline offers a structured framework for conducting an interactive and insightful discussion with young adults on the topic of environmental sustainability. Facilitators can customize the guideline to align with the research objectives, participants' characteristics, and the specific focus of the FGD. The well-organized structure ensures that the discussion covers various aspects of the research topic and encourages participants to share their thoughts and experiences openly.]

Source: Author

5.5 Sample Guideline for In-depth Case Interviews

Introduction:

Researcher: Good [morning/afternoon/evening], my name is [Your Name], and I'm conducting this interview as part of a research project aimed at gaining insights into the experiences of domestic violence among husband and wife garment worker couples. Your participation is incredibly valuable, and your privacy will be fully respected. Before we begin, please remember that you can skip any questions that you do not wish to answer, and your responses will remain confidential. Do you have any questions or concerns before we start?

Background Information:

Participant Information:
Participant 1 (Wife):

Name (Optional):
Age:
Length of Service in the Garment Industry:
Length of Service in the Garment Industry:

Section 1: Domestic Violence Experiences

2. Can you describe any instances of domestic violence within your relationship?

- Encourage participants to share specific incidents, including when, where, and how they occurred.
- When did these experiences of domestic violence begin, and have they evolved?

Section 2: Impact on Relationship and Work

4. How has domestic violence affected your relationship as husband and wife?

- Explore changes in communication, trust, emotional well-being, and family dynamics.
- Have there been any noticeable impacts on your work in the garment industry due to domestic violence?
- Consider absenteeism, productivity, concentration, or interactions with colleagues.

Section 3: Coping and Seeking Help

6. What strategies or mechanisms have you employed to cope with domestic violence within your relationship?
Discuss personal coping strategies, support from friends or family, or any formal interventions.
Have either of you sought support or assistance from your workplace or colleagues regarding these domestic violence issues?

Section 4: Legal and Reporting

8. Have you ever considered reporting domestic violence incidents to the authorities or management within your workplace?
Explore the factors influencing their decisions regarding reporting.

What challenges or barriers have you encountered when contemplating reporting domestic violence?

Section 5: Support Services

10. Are you aware of any support services or resources provided by your workplace or the local community to assist individuals facing domestic violence?

   - Include knowledge about counseling services, helplines, or shelters.

11. Have either of you utilized any of these support services, and if so, how helpful were they?

Section 6: Collective Action and Awareness

12. Do you believe there is sufficient awareness and support for husband and wife garment worker couples experiencing domestic violence within your community?

   - Consider community initiatives, awareness campaigns, or collective efforts.

13. In your opinion, what actions or initiatives could be taken to address domestic violence more effectively within the garment worker community?

Conclusion:

14. Is there anything else you would like to share about your experiences with domestic violence, any additional insights, or specific challenges faced as husband and wife garment worker couples?

Closure:

Researcher: Thank you very much for participating in this interview. Your willingness to share your experiences and perspectives is crucial to our research. Your well-being is important to us, and we genuinely appreciate your time and contribution. If you ever feel the need to talk further or seek support after this interview, please don't hesitate to reach out. We wish you the very best.

[Remember to approach this interview with sensitivity, empathy, and respect for the participants' experiences. Provide them with resources for support, and ensure their comfort throughout the research process.]

Source: Author

5.6 Sample Guideline for Participatory Rural Appraisal (PRA)

Participatory Rural Appraisal (PRA) is a participatory research approach that involves community members in assessing and analyzing their own development needs and priorities. A well-structured PRA guideline ensures that community members actively contribute and collaborate in the assessment process. Below is a sample guideline for conducting a PRA session focused on identifying livelihood challenges in a rural community.

Participatory Rural Appraisal (PRA) Guideline

Research Title: Identifying Livelihood Challenges in XYZ Rural Community

Introduction:

- Welcome participants and explain the purpose of the PRA session.
- Emphasize the importance of community members' knowledge and experiences.

Objective: To collaboratively identify and prioritize livelihood challenges faced by community members in the XYZ rural community.

PRA Session Setting: Community Hall, XYZ Rural Community

Date and Time: [Date and Time of PRA Session]

Duration: [Duration of PRA Session]

Facilitator: [Name of Facilitator]

Recording: Explain that the PRA session will be documented for research purposes.

Guideline:

- Icebreaker: Begin with an icebreaker activity to help participants become comfortable and build rapport.
- Introduce the PRA process: Explain the concepts of participation, empowerment, and collaborative analysis.
- Share the research objectives: Briefly describe the aim of identifying livelihood challenges and how the community's insights will contribute to the research.

Mapping Exercise:

- Provide a large map of the community.
- Ask participants to identify important locations, resources, and landmarks (e.g., markets, schools, water sources).
- Discuss the significance of these locations in livelihoods.

Seasonal Calendar:

- Create a calendar on a large sheet of paper.
- Ask participants to identify and mark seasons, agricultural activities, and events that impact their livelihoods.

Matrix Ranking:

- Present a list of potential livelihood challenges on cards.
- Ask participants to collectively rank these challenges based on their perceived significance.
- Discuss the reasons behind their rankings.

Daily Activity Analysis:

- Invite participants to describe their typical daily activities related to livelihoods.
- Document activities and the time spent on each.

Problem Tree:

- Draw a tree on a large sheet of paper.
- Ask participants to identify and discuss the root causes (problems) leading to the identified livelihood challenges (leaves).

Community Solutions:

- Engage participants in brainstorming solutions to address the identified livelihood challenges.
- Encourage creative thinking and suggestions.

Reflection and Sharing:

- Invite participants to share their insights and reflections on the PRA process.
- Discuss the value of collaborative analysis and the potential for positive change.

Conclusion:

- Express gratitude to the participants for their active participation.
- Explain the next steps, including how the insights gathered will contribute to further research and community development efforts.

[This Participatory Rural Appraisal (PRA) guideline provides a structured framework for conducting a collaborative assessment of livelihood challenges in a rural community. Facilitators can adapt and customize the guidelines to fit the research objectives, community context, and specific livelihood challenges being addressed. The interactive and participatory nature of PRA ensures that community members are actively involved in the assessment process and contribute their valuable insights and perspectives.]

Source: Author

5.7 Sample Guideline for Discourse Analysis

Discourse analysis is a research method used to study language use, communication patterns, and underlying social dynamics within texts or spoken interactions. A well-structured discourse analysis guideline ensures systematic analysis and interpretation of discourse content. Below is a sample guideline for conducting discourse analysis on media coverage of environmental issues.

Discourse Analysis Guideline

Research Title: Discourse Analysis of Media Coverage on Environmental Issues

Introduction:

- Introduce yourself and explain the purpose of the discourse analysis.
- Provide an overview of the research objectives and the focus on media coverage of environmental issues.

Objective: To analyze media discourse surrounding environmental issues and explore underlying narratives, frames, and perspectives.

Discourse Analysis Setting: Research Workspace

Date of Analysis: [Date of Analysis]

Duration of Analysis: [Duration of Analysis]

Analyst: [Name of Analyst]

Recording and Documentation: Explain that the analysis process will be documented for research documentation and reporting.

Guideline:

Data Collection:

- Gather a diverse set of media texts (news articles, opinion pieces, editorials) related to environmental issues.
- Organize the texts by theme, source, and date.

Familiarization:

- Read through the selected texts to become familiar with their content, context, and tone.
- Note initial impressions and emerging themes.

Data Coding:

- Develop a coding framework based on key themes and concepts related to environmental issues.
- Create a codebook that outlines definitions and guidelines for each code.

Open Coding:

- Begin the analysis process by reading each text line by line.
- Apply open codes to capture recurring words, phrases, and ideas related to environmental issues.

Developing Categories:

- Review the open codes and identify common patterns and categories.
- Group similar codes into overarching categories (e.g., climate change discourse, conservation discourse).

Thematic Analysis:

- Explore the relationships between categories and themes.
- Identify dominant narratives, frames, and perspectives within the discourse.

Language and Rhetoric:

- Analyze linguistic choices, rhetorical devices, and metaphors used in the discourse.
- Examine how language contributes to the shaping of discourse narratives.

Power Dynamics: Investigate power dynamics within the discourse, including who has a voice and whose perspectives are marginalized.

Contextual Analysis:

- Consider the broader socio-political context in which the discourse occurs.
- Reflect on how external factors influence the framing of environmental issues.

Interpretation and Findings:

- Reflect on the insights gained from the analysis.
- Develop an interpretive narrative that presents the identified discourse patterns, narratives, and perspectives.

Conclusion:

- Summarize the key findings of the discourse analysis.
- Reflect on the implications of the discourse patterns for public understanding of environmental issues.

[This discourse analysis guideline provides a structured framework for conducting an in-depth analysis of media coverage on environmental issues. Analysts can adapt and customize the guidelines to align with the research objectives, specific environmental issues, and the media texts being analyzed. The systematic approach ensures that discourse content is critically examined to uncover underlying narratives, frames, and perspectives shaping public discourse on environmental topics.]

Source: Author

5.8 Sample Checklist for Key Informant Interviews (KIIs)

Key Informant Interviews (KIIs) involve in-depth conversations with individuals who possess specialized knowledge and insights relevant to the research topic. A checklist for KIIs ensures that important aspects are covered during the interviews and that the necessary information is gathered. Below is a sample checklist for conducting KIIs on the theme of "Challenges and Opportunities in Sustainable Agriculture."

Key Informant Interview (KII) Checklist

Research Title: Exploring Challenges and Opportunities in Sustainable Agriculture

Interview Date: [Date of Interview]

Interviewer: [Name of Interviewer]

Key Informant: [Name of Key Informant]

Objective: To gather insights and expert opinions on challenges and opportunities in sustainable agriculture.

Interview Setting: [Location of Interview]

Duration of Interview: [Duration of Interview]

Recording: Explain that the interview will be audio-recorded for accurate documentation.

Checklist:

1. Introduction:

- ☐ Greet the key informant and introduce yourself.
- ☐ Explain the research objectives and the importance of their insights.

2. Background and Expertise:

- ☐ Ask about the key informant's background, expertise, and experience in sustainable agriculture.

3. Challenges in Sustainable Agriculture:

- ☐ Discuss challenges faced by farmers in adopting sustainable agricultural practices.
- ☐ Explore factors that hinder the implementation of sustainable methods.

4. Opportunities and Innovations:

- ☐ Inquire about innovative practices or technologies that offer opportunities for sustainable agriculture.
- ☐ Discuss the potential benefits of adopting these innovations.

5. Community and Stakeholder Engagement:

- ☐ Ask about the role of communities and stakeholders in promoting sustainable agriculture.
- ☐ Explore collaborative efforts and partnerships that contribute to sustainable practices.

6. Policy and Support:

- □ Discuss the role of government policies and incentives in encouraging sustainable agriculture.
- □ Inquire about support systems available to farmers for transitioning to sustainable methods.

7. Success Stories:

- □ Request examples of successful sustainable agriculture projects or initiatives.
- □ Discuss the factors that contributed to their success.

8. Barriers to Change:

- □ Inquire about potential barriers to adopting sustainable practices, such as economic challenges or lack of awareness.

9. Future Outlook:

- □ Ask about the key informant's vision for the future of sustainable agriculture.
- □ Discuss the potential impact of sustainable practices on food security and the environment.

10. Conclusion:

- □ Thank the key informant for their valuable insights.
- □ Ask if there are any additional points or perspectives they would like to share.

Follow-up Steps:

- □ Express interest in maintaining contact for potential further discussions or clarifications.
- □ Provide information about the research progress and outcomes.

[This Key Informant Interview (KII) checklist provides a structured framework for conducting in-depth interviews with experts in the field of sustainable agriculture. Researchers can adapt the checklist to align with the research objectives and the specific expertise of the key informants. The systematic approach ensures that important topics related to challenges, opportunities, and perspectives on sustainable agriculture are covered during the interview.]

Source: Author

5.9 Sample Checklist for Community Mapping

Community mapping is a participatory method that involves community members visually representing their local environment, resources, and important features. A checklist for community mapping ensures that the process is organized, inclusive, and captures key aspects of the community's landscape. Below is a sample checklist for conducting a community mapping exercise focused on "Mapping Community Resources and Services."

Community Mapping Checklist

Research Title: Mapping Community Resources and Services

Mapping Date: [Date of Mapping Exercise]

Facilitator: [Name of Facilitator]

Objective: To collaboratively map community resources, services, and important features to gain insights into the local environment.

Mapping Location: Community Center, [Community Name]

Duration of Mapping Exercise: [Duration of Mapping]

Recording: Explain that the mapping exercise will be documented for research documentation and reporting.

Checklist:

1. Introduction:

- Welcome participants and introduce yourself as the facilitator.
- Explain the purpose of the mapping exercise and how it will contribute to the research.

2. Mapping Tools and Materials:

- Set up tables, markers, pens, paper, and any other materials needed for the mapping activity.
- Ensure that participants have access to all required tools.

3. Base Map:

- Provide a base map of the community area, including streets, landmarks, and key features.
- Explain the layout of the base map to participants.

4. Identifying Resources and Services:

- Encourage participants to identify and mark important resources, services, and features on the map.
- Guide participants to add symbols, labels, and descriptions to their markings.

5. Categories of Mapping:

- Health Services
- Education Facilities
- Recreational Areas
- Markets and Shops
- Transportation Hubs
- Water Sources
- Community Spaces
- Religious Institutions

6. Group Discussions:

- Facilitate discussions as participants share their insights and reasoning for mapping specific resources.
- Encourage participants to ask questions and exchange information.

7. Representation of Features:

- Ensure that resources and features are accurately represented in terms of location and size on the map.

8. Inclusivity:

- Ensure that all participants have an opportunity to contribute and that no voice dominates the discussion.

9. Reflective Discussions:

- Pause periodically to discuss the patterns and connections emerging from the mapping activity.
- Encourage participants to reflect on the significance of the mapped resources and services.

10. Presentation and Discussion:

- Invite participants to present their mapped areas and share insights with the entire group.
- Facilitate a group discussion on common observations and shared perspectives.

11. Documentation:

- Document the mapped areas, symbols, and discussions using photographs or written notes.

Conclusion:

- Thank participants for their active participation and contributions.
- Explain how the mapped information will be used for research purposes and community development.

[This Community Mapping checklist offers a structured framework for conducting a participatory mapping exercise to identify and document community resources and services. Facilitators can adapt and customize the checklist to fit the research objectives, community context, and specific categories of resources and features being mapped. The inclusive and interactive nature of community mapping ensures that local knowledge is captured and valuable insights are gained into the community's environment and services.]

Source: Author

5.10 Sample Checklist for Secondary Sources

Secondary sources encompass a wide range of existing materials, such as academic papers, reports, books, and articles, that are analyzed to provide insights and support for research. A checklist for analyzing secondary sources ensures that relevant information is systematically extracted and critically evaluated. Below is a sample checklist for analyzing secondary sources related to "Global Trends in Renewable Energy Adoption."

Secondary Source Analysis Checklist

Research Title: Global Trends in Renewable Energy Adoption

Analysis Date: [Date of Secondary Source Analysis]

Analyst: [Name of Analyst]

Objective: To analyze and extract relevant insights from secondary sources regarding global trends in renewable energy adoption.

Checklist:

1. Source Identification:

- Document the title, author(s), publication date, and source of each secondary document.

2. Research Objectives:

- Review the research objectives and questions to guide the selection of relevant secondary sources.

3. Literature Review:

- Conduct a comprehensive literature review to identify a diverse range of secondary sources on the chosen topic.

4. Source Evaluation:

- Evaluate the credibility, authority, and relevance of each secondary source.

- Consider the source's publication outlet, the reputation of the author, and the recency of publication.

5. Data Extraction:

- Extract key data, statistics, findings, and arguments from each secondary source.
- Note relevant quotations, graphs, tables, and case studies.

6. Categorization:

- Categorize secondary sources based on themes, trends, or subtopics relevant to renewable energy adoption.

7. Comparative Analysis:

- Analyze and compare findings, trends, and perspectives across multiple secondary sources.
- Identify areas of consensus and divergence among different sources.

8. Cross-Referencing:

- Cross-reference information from different sources to ensure accuracy and reliability.

9. Critique and Interpretation:

- Critically evaluate the strengths and limitations of each secondary source.
- Interpret the data and findings within the context of the research objectives.

10. Identification of Gaps:

- Identify gaps or areas where existing secondary sources may lack comprehensive coverage.

11. Synthesis of Insights:

- Synthesize the extracted insights and findings from the secondary sources.
- Summarize key trends, challenges, opportunities, and emerging patterns.

12. Citations and References:

- Ensure proper citation of secondary sources using the appropriate citation style.

13. Ethical Considerations:

- Ensure ethical handling of secondary sources by providing proper attribution and avoiding plagiarism.

Conclusion:

- Summarize the main insights gained from the analysis of secondary sources.
- Reflect on the contribution of secondary sources to the research objectives.

[This Secondary Source Analysis checklist provides a structured framework for systematically analyzing and extracting insights from a diverse range of secondary sources. Analysts can adapt and customize the checklist to align with the research objectives, specific research topics, and the types of secondary sources being analyzed. The systematic approach ensures that relevant information is gathered, evaluated, and synthesized to contribute to the overall research analysis and findings.]

Source: Author

CHAPTER 6

Using Technology in Data Collection

ABSTRACT

This chapter explores the seamless integration of technological advancements into the research process. It delves into the integration of technology in research, the utilization of online surveys and digital questionnaires, and the efficiency gained through mobile apps for data collection. This chapter illuminates how technology enhances the effectiveness and precision of data collection in contemporary research.

Keywords: Technology integration, research tools, online surveys, digital questionnaires, mobile apps, data collection efficiency

6.1 Integration of Technology in Research

Technology has revolutionized the field of research by offering innovative tools and solutions that enhance data collection processes. The integration of technology in research enables researchers to collect, analyze, and manage data more efficiently, accurately, and in real-time. Below is an overview of how technology is integrated into research and the benefits it brings to the data collection process.

6.1.1 Key Areas of Integration of Technology in Research

Technology has transformed the research landscape, offering a wide array of tools and platforms that streamline data collection, analysis, and management. The integration of technology in research encompasses various stages of the research process, from designing data collection instruments to analyzing and presenting findings. Key areas where technology plays a crucial role include:

1. Data Collection Instruments: Technology allows researchers to design digital data collection instruments such as online surveys, mobile apps, and electronic questionnaires. These instruments offer several advantages, including automated data entry, reduced errors, and the ability to include multimedia elements like images and videos. Online surveys and mobile apps also enable data collection from a geographically dispersed audience.
2. Remote Data Collection: Advancements in technology enable researchers to collect data remotely. Mobile devices and online platforms facilitate remote interviews, surveys, and observations. This approach is particularly valuable for collecting data from hard-to-reach populations or during times when physical presence is limited.
3. Sensors and IoT Devices: Researchers can integrate sensors and Internet of Things (IoT) devices to gather real-time data. These devices can collect environmental, health-related, or other data continuously, providing a wealth of information that would be challenging to capture using traditional methods.
4. Big Data Analysis: Technology supports the analysis of large datasets through advanced data analytics tools. Researchers can uncover patterns, trends, and correlations that might go unnoticed in smaller datasets. Big data analysis is especially relevant in fields like healthcare, finance, and social sciences.
5. Data Management Systems: Digital platforms and software systems facilitate efficient data management and organization. Researchers can store, sort, and access data securely, making it easier to track and manage complex datasets.
6. Visualization Tools: Technology offers visualization tools that enable researchers to create graphs, charts, and interactive visual

representations of data. These visualizations aid in presenting complex information in an easily understandable format.

7. Cloud Computing: Cloud computing services provide scalable storage and computing resources. Researchers can securely store and process data without the need for extensive infrastructure investments.
8. Collaboration and Sharing: Technology allows for seamless collaboration among researchers regardless of their geographical locations. Online platforms enable team members to work on the same project simultaneously, share data, and provide real-time feedback.

6.1.2 Benefits of Integrating Technology

The integration of technology in research brings several benefits that enhance the data collection process and research outcomes:

Efficiency: Technology streamlines data collection, reducing manual entry and the potential for errors.

Accuracy: Automated processes minimize human errors and transcription mistakes.

Real-time Insights: Researchers can access and analyze data in real-time, enabling quicker decision-making.

Global Reach: Technology facilitates data collection from diverse geographical locations and populations.

Flexibility: Remote data collection and digital instruments offer flexibility in reaching participants.

Cost-Effectiveness: Digital tools often eliminate the need for printing, distribution, and manual data entry.

Data Security: Digital data management systems offer enhanced security and backup options.

Rich Data: Technology allows for the collection of multimedia-rich data, including images, audio, and video.

Incorporating technology into the research process requires careful consideration of factors such as data privacy, participant consent, and the accessibility of digital tools to all participants. However, when used thoughtfully, technology can greatly enhance the efficiency, accuracy, and depth of insights obtained during data collection, contributing to more robust research outcomes.

6.2 Online Surveys and Digital Questionnaires

Online surveys and digital questionnaires have become essential tools in modern research, offering a convenient and efficient way to collect data from a wide range of participants. These digital methods provide researchers with the flexibility to reach diverse populations, gather real-time responses, and analyze data more effectively. Here's an overview of online surveys and digital questionnaires and their benefits in the data collection process.

6.2.1 Definition of Online Surveys

Online surveys are web-based questionnaires that participants can complete through internet-connected devices such as computers, smartphones, or tablets. Researchers design the survey using online survey platforms and distribute it to participants electronically. Participants respond to the questions by selecting options, providing text-based answers, or rating items on a scale.

6.2.2 Benefits of Online Surveys

Accessibility and Convenience: Participants can access and complete surveys from anywhere, eliminating geographical constraints and allowing for a larger and more diverse sample.

Cost-Effective: Online surveys eliminate printing and postage costs associated with traditional paper surveys.

Real-Time Data: Researchers can access responses in real-time, enabling immediate data analysis and decision-making.

Automated Data Collection: Responses are automatically collected and recorded, reducing the need for manual data entry and transcription.

Quick Dissemination: Surveys can be distributed to a large number of participants simultaneously, saving time and effort in data collection.

Customization: Researchers can customize the appearance and structure of the survey, including branching logic based on participant responses.

Data Validation: Online surveys can include validation rules to ensure accurate and consistent responses.

6.2.3 Digital Questionnaires

Digital questionnaires are electronic versions of traditional paper questionnaires. They are designed using software applications and can be distributed via email or shared through online platforms. Participants fill out the questionnaire digitally, and the responses are automatically captured and recorded in digital format.

Benefits of Digital Questionnaires:

Reduced Paper Usage: Digital questionnaires eliminate the need for printing and physical distribution, contributing to environmental sustainability.

Efficiency: Automated data capture reduces errors and accelerates the data collection process.

Flexibility: Participants can complete digital questionnaires at their convenience, enabling a more flexible and accommodating data collection process.

Data Integrity: Responses are digitally recorded, minimizing the risk of data loss or damage.

Integration with Analysis Tools: Digital responses can be easily imported into data analysis software for streamlined analysis.

Remote Participation: Participants can engage in data collection regardless of their location, increasing the reach of the research.

Engagement: Digital questionnaires can incorporate multimedia elements such as images or videos to enhance participant engagement.

Considerations and Best Practices:

When using online surveys and digital questionnaires, researchers should consider factors such as data security, participant privacy, and accessibility. It's important to:

- Ensure data encryption and secure storage of responses.
- Obtain informed consent from participants before they begin the survey.
- Design user-friendly and responsive surveys that work well on various devices.
- Pilot test the survey to identify any technical issues or confusing questions.
- Provide clear instructions and guidance for participants.
- Consider offering incentives to encourage participation.

Online surveys and digital questionnaires offer researchers a powerful means to collect data efficiently, engage participants remotely, and access real-time insights. By embracing these digital methods, researchers can enhance the quality and reach of their data collection efforts, leading to more robust research outcomes.

6.3 Mobile Apps and Data Collection Efficiency

Mobile apps have become powerful tools for researchers to streamline data collection processes, enhance participant engagement, and improve the overall efficiency of data collection. These apps offer a range of features that enable researchers to gather data in real-time, ensure data accuracy, and access insights more quickly. Here, we explore the benefits of using mobile apps for data collection and their impact on research efficiency.

6.3.1 Benefits of Using Mobile Apps for Data Collection:

Real-Time Data Collection: Mobile apps enable researchers to collect data in real-time, eliminating delays associated with manual data entry and transcription.

Increased Accuracy: Digital data collection reduces the risk of human errors, as participants input responses directly into the app, reducing the need for data interpretation.

Multimedia Integration: Mobile apps allow participants to capture and attach multimedia elements such as photos, videos, and audio recordings to their responses, enriching the data.

Geo-Tagging: Many mobile apps have GPS capabilities, allowing researchers to capture location-specific data, which is particularly useful for studies involving geography or environmental data.

Offline Data Collection: Some mobile apps offer offline data collection options, allowing participants to complete surveys or forms without an active internet connection.

Customization: Researchers can design and customize surveys or data collection forms to match their specific research objectives, including skip logic and branching questions.

Efficient Data Management: Mobile apps automatically store collected data in digital format, reducing the need for manual data entry and organization.

Participant Engagement: Interactive features of mobile apps, such as touch-based interactions and multimedia content, enhance participant engagement and reduce survey fatigue.

Data Security: Many mobile apps provide data encryption and secure cloud storage, ensuring participant privacy and data security.

Time and Cost Savings: Mobile apps eliminate the need for paper-based materials, printing, and manual data entry, resulting in cost and time savings.

6.3.2 Impact on Research Efficiency

The integration of mobile apps into data collection processes significantly enhances research efficiency in several ways:

Faster Data Collection: Mobile apps facilitate the real-time collection of data, allowing researchers to gather responses quickly and immediately access insights.

Reduced Errors: Mobile apps reduce the potential for errors that may occur during manual data entry, transcription, or data transfer.

Automated Analysis: Data collected through mobile apps can often be exported in formats compatible with data analysis tools, enabling immediate analysis.

Rapid Reporting: Researchers can generate preliminary reports and summaries in real-time, expediting the dissemination of key findings.

Flexibility: Researchers can adapt the data collection process on the fly, making adjustments to survey questions or methods as needed.

Remote Access: Researchers can access data remotely and monitor data collection progress in real-time, even if they are not physically present.

6.3.3 Considerations and Best Practices

While mobile apps offer numerous benefits for data collection, researchers should consider the following best practices:

- Select an app that aligns with research goals and target audience.
- Pilot test the app to ensure its functionality and usability.

- Prioritize participant privacy and informed consent.
- Offer technical support to participants if needed.
- Keep surveys concise and user-friendly for mobile devices.
- Verify data synchronization and backup mechanisms.
- Plan for data analysis and reporting tools that integrate with collected data.

By leveraging the capabilities of mobile apps, researchers can enhance data collection efficiency, improve data quality, and accelerate the research process. Mobile apps provide a platform for engaging participants, collecting accurate data, and gaining insights in real-time, ultimately contributing to more effective research outcomes.

CHAPTER 7

Scaling and Validating Instruments

ABSTRACT
This chapter provides into the critical process of ensuring the reliability and validity of data collection instruments. It explores methods to guarantee measurement validity and reliability, the importance of pilot testing, and strategies for preparing instruments for scale. Additionally, this chapter introduces statistical approaches that researchers can employ to validate their instruments effectively.

Keywords: Measurement validity, reliability, pilot testing, instrument scaling, statistical validation, research instruments

7.1 Ensuring Measurement Validity and Reliability

In research, measurement validity and reliability are critical aspects that ensure the accuracy and credibility of collected data. Validity refers to the extent to which an instrument measures what it is intended to measure, while reliability refers to the consistency and stability of measurement over time and across different conditions. Ensuring high levels of validity and reliability is essential to drawing accurate conclusions from collected data. Here, we delve into the concepts of measurement validity and reliability and discuss strategies to enhance them.

7.1.1 Measurement Validity

Measurement validity addresses whether a measurement instrument accurately assesses the construct it is intended to measure. There are several types of validity to consider:

Content Validity: The extent to which the items in a measurement instrument comprehensively cover the domain of the construct.

Construct Validity: The degree to which a measurement instrument effectively measures the underlying theoretical construct it is intended to represent.

Criterion Validity: The extent to which scores on a measurement instrument correlate with external criteria that are theoretically related to the construct.

Convergent Validity: The degree to which scores on a measurement instrument correlate with other measures of the same construct.

Discriminant Validity: The degree to which scores on a measurement instrument do not correlate with measures of unrelated constructs.

7.1.2 Measurement Reliability

Measurement reliability refers to the consistency and stability of the results produced by a measurement instrument. It is assessed through methods such as:

Test-Retest Reliability: Consistency of scores when the same instrument is administered to the same participants at different points in time.

Internal Consistency Reliability: Consistency of scores across different items within the same measurement instrument.

Inter-Rater Reliability: Consistency of scores when the same measurement instrument is administered by different raters or observers.

7.1.3 Strategies for Enhancing Validity and Reliability

Pilot Testing: Conduct pilot testing to identify any ambiguities, inconsistencies, or difficulties in understanding the measurement instrument by participants. Adjustments can be made based on participant feedback.

Expert Review: Seek input from experts in the field to evaluate the content validity of the measurement instrument. Experts can provide insights into the relevance and comprehensiveness of the items.

Construct Validation: Administer the measurement instrument alongside established measures of the same construct to assess convergent and discriminant validity.

Criterion Validation: Correlate scores on the measurement instrument with external criteria that are theoretically related to the construct being measured.

Randomization: If applicable, randomize the order of items in the measurement instrument to minimize order effects.

Test-Retest: Administer the measurement instrument to a subset of participants on two separate occasions to assess test-retest reliability.

Inter-Rater Consistency: If relevant, conduct inter-rater reliability assessments by having multiple raters independently score the same responses.

Internal Consistency: Calculate measures of internal consistency such as Cronbach's alpha for items within the same measurement instrument.

Refinement: Continuously refine the measurement instrument based on the insights gained from pilot testing, expert review, and psychometric analyzes.

Adaptation: When using established measurement instruments, ensure they are culturally and contextually adapted to maintain validity and reliability.

Ensuring high validity and reliability contributes to the credibility of research findings and enhances the overall quality of the study. Researchers should adopt a rigorous approach to developing and validating measurement instruments, as this foundation is essential for drawing accurate conclusions and making meaningful interpretations based on collected data.

7.2 Pilot Testing and Preparing Instruments for Scale

Before implementing data collection on a larger scale, pilot testing plays a crucial role in identifying and addressing potential issues with measurement instruments. Pilot testing involves administering the instruments to a small group of participants to evaluate their clarity, effectiveness, and the overall data collection process. Successful pilot testing ensures that the instruments are ready for wider use and helps researchers refine their approach. Here, we delve into the significance of pilot testing and the steps involved in preparing instruments for larger-scale data collection.

7.2.1 Significance of Pilot Testing

Identifying Ambiguities: Pilot testing reveals any ambiguities, confusing wording, or inconsistencies in the measurement instruments that could lead to incorrect responses.

Understanding Participant Experience: Researchers gain insights into how participants perceive and respond to the questions, helping to ensure that the instruments are user-friendly and comprehensible.

Addressing Technical Issues: Pilot testing helps identify technical glitches, such as errors in online surveys or mobile apps, that could impact data collection.

Testing Data Collection Process: The pilot test mimics the actual data collection process, allowing researchers to assess the logistics, timing, and overall feasibility of the process.

7.2.2 Steps for Pilot Testing and Instrument Preparation

Select a Small Sample: Choose a small, diverse sample of participants who represent the intended population. This sample should be separate from the main study group.

Administer the Instruments: Administer the measurement instruments to participants as if it were the actual data collection phase. This includes any technology or methods to be used.

Gather Feedback: Collect feedback from participants about the clarity, relevance, and comprehensibility of the questions. Use open-ended questions to capture qualitative feedback.

Analyze Data: Analyze the pilot test data to identify any patterns of confusion, inconsistent responses, or technical issues.

Refine the Instruments: Based on the feedback and data analysis, make necessary revisions to the measurement instruments to address any issues or shortcomings.

Expert Review: Seek input from experts or colleagues in the field to review the revised instruments for content validity and relevance.

Repeat Pilot Testing (if needed): If significant changes were made, consider repeating the pilot testing with a new small sample to ensure that the revisions have resolved the issues.

Finalize Instruments: Once the measurement instruments have undergone revisions and refinements based on pilot testing and expert input, finalize them for use in the larger-scale data collection.

Communication: Communicate any changes or updates to the data collection process and instruments to the research team and participants.

Training: If necessary, provide training to data collectors to ensure they understand the instruments, procedures, and any changes that have been made.

By conducting thorough pilot testing and preparing instruments for scale, researchers can ensure that their data collection process is efficient, reliable, and accurate. Pilot testing allows for the identification and resolution of potential issues early in the research process, ultimately contributing to the success of the larger-scale data collection and the validity of research findings.

7.3 Statistical Approaches to Validation

Statistical approaches are essential for assessing and validating the measurement instruments used in research. These approaches help researchers quantify the extent to which their instruments measure the intended constructs accurately and consistently. By employing various statistical techniques, researchers can provide evidence of the validity and reliability of their measurement instruments. Here, we explore some of the key statistical methods used in validation:

1. Confirmatory Factor Analysis (CFA): CFA is a statistical technique used to assess the construct validity of measurement instruments. It involves testing a hypothesized measurement model to determine whether the observed variables (items) load onto the latent constructs as expected. CFA helps researchers evaluate how well the measurement instrument reflects the underlying theoretical construct.
2. Exploratory Factor Analysis (EFA): EFA is used to explore the underlying factor structure of a set of observed variables. It is particularly useful when researchers are unsure about the specific relationships between items and constructs. EFA helps identify patterns of item loadings and can guide the development of a valid measurement model.
3. Cronbach's Alpha: Cronbach's alpha is a commonly used measure of internal consistency reliability. It assesses the extent to which the items within a measurement instrument consistently measure the same construct. Higher values of Cronbach's alpha indicate stronger internal consistency and reliability.

4. Test-Retest Reliability: Test-retest reliability assesses the consistency of measurement instrument scores over time. Researchers administer the same instrument to the same participants on two separate occasions and calculate the correlation between the two sets of scores. A high correlation indicates good test-retest reliability.

5. Inter-Rater Reliability: Inter-rater reliability measures the consistency of scores when the same measurement instrument is administered by different raters or observers. It is commonly used in research involving observational or qualitative data.

6. Pearson Correlation Coefficient: Pearson correlation coefficient assesses the strength and direction of the linear relationship between two continuous variables. It can be used to evaluate the criterion validity of a measurement instrument by comparing its scores with external criteria.

7. Convergent and Discriminant Validity: Convergent validity is assessed by examining the correlations between a measurement instrument and other measures of the same construct. Discriminant validity is assessed by comparing the correlations between a measurement instrument and measures of unrelated constructs. These analyzes demonstrate that the instrument correlates more strongly with similar constructs than with dissimilar ones.

8. Item Analysis: Item analysis involves evaluating the performance of individual items within a measurement instrument. Researchers assess factors such as item difficulty, item discrimination, and item-total correlations to identify problematic items.

9. Factorial Validity: Factorial validity examines how well the observed items align with the expected factors or dimensions within a construct. It involves analyzing the pattern of item loadings in factor analysis.

10. Sensitivity and Specificity Analysis: Sensitivity and specificity analysis are relevant when the measurement instrument is used to identify or classify participants based on a certain criterion. Sensitivity measures the instrument's ability to correctly identify positive cases, while specificity measures its ability to correctly identify negative cases.

Employing appropriate statistical approaches enhances the credibility of research findings by providing evidence of the validity and reliability of measurement instruments. Researchers should select the most suitable statistical techniques based on the nature of their research, the type of data collected, and the intended constructs of measurement.

CHAPTER 8

Ethical Considerations in Data Collection

ABSTRACT
This chapter explores the ethical dimensions' integral to responsible research practice. It emphasizes the importance of respecting privacy and confidentiality, navigating the complexities of informed consent and participant rights, and addressing ethical dilemmas that may arise during data collection. This chapter provides researchers with a moral compass to uphold the highest ethical standards in their work.

Keywords: Ethical considerations, privacy, confidentiality, informed consent, participant rights, ethical dilemmas, research ethics

8.1 Respecting Privacy and Confidentiality

Ethical considerations are paramount in research, especially when collecting data from participants. Respecting privacy and confidentiality ensures that participants' rights are protected and builds trust between researchers and participants. Here, we explore the significance of privacy and confidentiality in data collection and provide guidelines for maintaining ethical standards.

8.1.1 Significance of Respecting Privacy and Confidentiality

Participant Trust: Respecting privacy and confidentiality fosters trust between researchers and participants, encouraging open and honest responses.

Participant Autonomy: Protecting privacy allows participants to share information willingly, maintaining their autonomy in deciding what to disclose.

Data Accuracy: When participants feel their privacy is safeguarded, they are more likely to provide accurate and truthful information.

Legal and Ethical Obligations: Many research ethics codes and regulations mandate the protection of participants' privacy and confidentiality.

8.1.2 Guidelines for Maintaining Privacy and Confidentiality

Informed Consent: Clearly explain the purpose of data collection, how the data will be used, and the measures in place to protect privacy. Obtain informed consent from participants before collecting any data.

Anonymity: Use anonymous data collection methods when feasible. Avoid collecting information that can directly identify participants, such as names or personal identification numbers.

Confidentiality: Assure participants that their responses will remain confidential. Store data securely and restrict access only to authorized personnel.

Secure Data Storage: Use secure methods to store collected data, such as encrypted databases or password-protected files. Limit access to individuals who need the data for analysis.

Data Sharing: If data sharing is necessary, de-identify data by removing personally identifiable information before sharing with others.

Pseudonyms: When reporting findings, use pseudonyms or code numbers to protect participants' identities.

Data Retention: Establish a clear policy for how long data will be retained and when it will be securely disposed of.

Sensitive Information: Handle sensitive information (e.g., medical history, personal beliefs) with extra care. Ensure participants' comfort and provide options to skip questions.

Communication: Maintain open communication with participants throughout the research process. Address any concerns they may have about privacy and confidentiality.

Data Access Requests: Establish a process for participants to access their data and request its removal if desired.

Ethics Review: Seek approval from an institutional review board (IRB) or ethics committee before initiating data collection. They ensure that privacy and confidentiality measures meet ethical standards.

8.1.3 Challenges and Considerations

Online Data Collection: Protect privacy when collecting data online by using secure platforms, encryption, and secure data transfer methods.

Vulnerable Populations: Take additional precautions when working with vulnerable populations, such as minors or individuals with diminished capacity, to ensure their privacy is safeguarded.

Cross-Cultural Sensitivity: Consider cultural norms and expectations around privacy, as they can vary across different societies.

Data Breach Preparedness: Have a plan in place to address potential data breaches, including informing affected participants promptly.

Data Sharing Agreements: If sharing data with other researchers or institutions, establish clear agreements outlining data usage, protection, and confidentiality.

By adhering to ethical principles of privacy and confidentiality, researchers uphold the rights of participants and maintain the integrity of their research. Ethical data collection practices not only protect participants but also contribute to the credibility and reputation of researchers and institutions.

8.2 Informed Consent and Participant Rights

Informed consent is a cornerstone of ethical research, ensuring that participants understand the research process, their role, and potential risks before agreeing to participate. Respecting participant rights and obtaining informed consent is crucial for upholding ethical standards in data collection. Here, we delve into the significance of informed consent and participant rights, along with guidelines for their implementation.

8.2.1 Significance of Informed Consent

Autonomy: Informed consent respects participants' autonomy, allowing them to make informed decisions about their involvement in research.

Transparency: It promotes transparency by providing participants with clear information about the study's purpose, procedures, potential risks, benefits, and their rights.

Legal and Ethical Obligations: Many research ethics codes and regulations mandate obtaining informed consent to protect participants' rights and well-being.

Trust: Informed consent builds trust between researchers and participants, as it demonstrates researchers' commitment to transparency and respect for participants' wishes.

8.2.2 Guidelines for Obtaining Informed Consent and Respecting Participant Rights

Clear and Understandable Language: Present all information in clear, simple language that participants can understand, avoiding jargon or technical terms.

Comprehensive Information: Provide comprehensive details about the research, including the purpose, procedures, risks, benefits, time commitment, and any compensation.

Voluntary Participation: Emphasize that participation is voluntary, and participants can withdraw at any time without consequences.

Privacy and Confidentiality: Explain how participant data will be protected, stored, and used, ensuring that privacy and confidentiality will be maintained.

Contact Information: Provide contact details for researchers, including names, affiliations, and contact information, so participants can reach out with questions or concerns.

Inclusion of Minors: When involving minors, obtain informed consent from both the minor and their legal guardian, ensuring their understanding and approval.

Vulnerable Populations: Take extra precautions when working with vulnerable populations, such as those with limited decision-making capacity, to ensure their informed consent is genuine and well-understood.

Time for Consideration: Allow participants sufficient time to read and comprehend the informed consent form before making a decision.

Opportunity for Questions: Provide a platform for participants to ask questions about the study, clarifying any doubts they may have.

Documentation: Obtain written or electronic consent from participants, and provide them with a copy of the informed consent form for their records.

Withdrawal: Clearly explain participants' right to withdraw from the study at any time without facing negative consequences.

Comprehension Assessment: Ensure participants understand the information by asking them to summarize key points in their own words.

Continuous Communication: Maintain open communication with participants throughout the study, addressing any concerns that may arise.

8.2.3 A Sample of Informed Consent Form

Creating an informed consent form is crucial for ensuring that research participants are fully aware of the nature, risks, and benefits of the study they are involved in. Here is a sample of an informed consent form that you can use as a template for your research projects. This example covers essential components including purpose, procedures, risks, benefits, confidentiality, and the participant's rights.

INFORMED CONSENT FORM

TITLE OF THE STUDY:...

PRINCIPAL INVESTIGATOR:.................................

INTRODUCTION

You are invited to participate in a research study conducted by .... from the .... This form provides information about the study to help you decide whether or not to participate. Please read this form carefully and feel free to ask any questions you may have.

PURPOSE OF THE STUDY

....

....

....

PROCEDURES

If you agree to participate in this study, you will be asked to:

- Complete a questionnaire that will take approximately . . . minutes.
- Participate in an interview lasting about . . . minutes, where you will discuss your experiences with data collection tools.
- Allow your responses to be recorded and analyzed for the purposes of this research.

RISKS AND DISCOMFORTS

There are minimal risks associated with this study. You may feel uncomfortable answering some of the questions. You can skip any question that makes you feel uncomfortable, and you may withdraw from the study at any time without any penalty.

BENEFITS

Your participation in this study will provide valuable insights into . . . . While there are no direct benefits to you, the findings may contribute to . . . ., which can benefit the broader community.

CONFIDENTIALITY

Your responses will be kept confidential. Data will be stored securely and will only be accessible to the research team. Any information that is published will not identify you by name. Your identity will remain anonymous in all reports and publications.

VOLUNTARY PARTICIPATION

Participation in this study is completely voluntary. You may refuse to participate or withdraw from the study at any time without any consequences.

CONTACT INFORMATION

If you have any questions about the study or your participation, please contact:

- **Name of Principal Investigator:**............................
- **Email:**..
- **Phone:**......................................

If you have questions about your rights as a research participant or wish to report a research-related issue, you may contact the.................. Institutional Review Board (IRB) at.................................

- **Email:**............................
- **Phone:**...........................

STATEMENT OF CONSENT

I have read and understood the information provided in this consent form. I have had the opportunity to ask questions and they have been answered to my satisfaction. I voluntarily agree to participate in this research study.

Participant's Name: ________________________________

Signature: ________________________________

Date: ________________________________

Researcher's Name: ________________________________

Signature: ________________________________

Date: ________________________________

This template provides a comprehensive framework for informed consent, ensuring that participants are fully aware of what the study involves and their rights as participants. You can modify the content to fit the specific needs of your study.

8.2.4 Challenges and Considerations

Cross-Cultural Sensitivity: Tailor informed consent processes to the cultural norms and expectations of the participants' communities.

Language Barriers: Provide consent forms in participants' preferred languages to ensure understanding.

Online Consent: When collecting data online, use secure platforms and mechanisms to obtain electronic consent.

Dynamic Consent: For longitudinal studies, maintain ongoing communication and ensure participants' understanding and consent as the study progresses.

Specialized Populations: For research involving clinical trials, medical interventions, or invasive procedures, ensure that participants fully comprehend potential risks and benefits.

By implementing robust informed consent processes and respecting participant rights, researchers demonstrate their commitment to ethical research practices and the well-being of those who contribute to their studies. Informed consent ensures that participants are well-informed, protected, and treated with respect throughout the research journey.

8.3 Navigating Ethical Dilemmas in Data Collection

Ethical dilemmas can arise unexpectedly during the data collection process, challenging researchers to make decisions that uphold ethical principles while balancing research goals. Navigating these dilemmas requires careful consideration, adherence to ethical guidelines, and a commitment to prioritizing participants' rights and well-being. Here, we explore common ethical dilemmas in data collection and strategies for addressing them.

8.3.1 Common Ethical Dilemmas in Data Collection

Informed Consent Challenges: Participants may struggle to fully comprehend the study's details, potentially leading to inadequate informed consent.

Privacy and Confidentiality Breaches: Maintaining privacy and confidentiality can become challenging when working with sensitive data or vulnerable populations.

Coercion and Undue Influence: Researchers must ensure that participants are not coerced or influenced to participate against their will.

Deception: Balancing the need for accurate responses and maintaining participant trust can be difficult when using techniques that involve deception.

Anonymity vs. Data Quality: Balancing the desire for anonymous data with the need for accurate responses can pose challenges, particularly in qualitative research.

Participant Well-Being: Researchers must ensure that participation does not harm participants physically, emotionally, or psychologically.

Dual Roles: Researchers may face dilemmas when their dual roles (e.g., researcher and therapist) create conflicts of interest.

8.3.2 Strategies for Addressing Ethical Dilemmas

Ethics Review: Seek guidance from an institutional review board (IRB) or ethics committee. They can provide insights into how to navigate ethical dilemmas while adhering to ethical standards.

Transparent Communication: Maintain open and transparent communication with participants about potential risks, benefits, and ethical considerations.

Informed Consent Process: Prioritize a robust informed consent process to ensure participants understand the study's details and potential implications.

Privacy Protection: Implement stringent data security measures to safeguard participants' privacy and prevent data breaches.

Participant Welfare: Prioritize participants' well-being and ensure that their physical and emotional safety is not compromised.

Debriefing: If deception was involved, provide participants with a thorough debriefing that explains the purpose and rationale for the deception.

Conflict of Interest Mitigation: Clearly define and communicate roles to avoid conflicts of interest, especially when dual roles are present.

Minimize Harm: Continuously assess the potential harm of research activities and take steps to minimize harm to participants.

Seek Consultation: Consult with colleagues, mentors, or ethics experts when facing complex ethical dilemmas to gain different perspectives.

Alternative Approaches: Consider alternative research methods or approaches that avoid ethical dilemmas altogether.

8.3.3 Ethical Decision-Making Process

Identify the Dilemma: Recognize the ethical dilemma and the conflicting ethical principles involved.

Gather Information: Collect all relevant information about the situation, including potential risks and benefits.

Evaluate Options: Consider various courses of action and their potential consequences.

Apply Ethical Guidelines: Refer to ethical guidelines, codes of conduct, and relevant regulations to inform your decision.

Make a Decision: Choose the course of action that best upholds ethical principles and prioritizes participants' rights and well-being.

Reflect and Learn: After addressing the dilemma, reflect on the experience and consider how to prevent similar dilemmas in the future.

Ethical dilemmas are a natural part of research, and navigating them requires thoughtful consideration, ethical awareness, and a commitment to doing what is right for participants and the research process. By actively engaging in ethical decision-making, researchers can uphold their professional integrity and ensure the ethical conduct of their research endeavors.

CHAPTER 9

Data Collection in Diverse Contexts

ABSTRACT

This chapter navigates the complexities of gathering data in a multicultural world. It underscores the significance of cultural sensitivity in instrument design, explores strategies for adapting tools for different demographics, and addresses cross-cultural challenges while providing practical solutions. This chapter equips researchers to conduct data collection with cultural awareness and competence.

Keywords: Cultural sensitivity, instrument design, demographic adaptation, cross-cultural challenges, cultural solutions, diverse contexts

9.1 Cultural Sensitivity in Instrument Design

Collecting data in diverse cultural contexts requires researchers to be sensitive to cultural norms, values, and beliefs. Cultural sensitivity ensures that the data collection process is respectful, relevant, and effective across different cultural groups. In this section, we explore the importance of cultural sensitivity in instrument design and provide guidelines for creating culturally appropriate data collection instruments.

9.1.1 Significance of Cultural Sensitivity in Instrument Design

Respect for Diversity: Cultural sensitivity acknowledges the diversity of cultural backgrounds and avoids imposing one cultural perspective on all participants.

Valid Data: Instruments designed with cultural sensitivity are more likely to elicit accurate and valid responses from participants who may have different linguistic, social, or cognitive frameworks.

Trust Building: Demonstrating cultural sensitivity builds trust between researchers and participants, enhancing their willingness to engage in the research process.

Accurate Interpretation: Culturally sensitive instruments enable researchers to accurately interpret and analyze data within the context of participants' cultural norms.

9.1.2 Guidelines for Culturally Sensitive Instrument Design

Cultural Consultation: Seek input from individuals who are familiar with the culture you are studying. Collaborate with local experts to ensure the instrument is appropriate and respectful.

Language and Translation: Use clear and culturally appropriate language in the instrument. If translating, ensure accurate translations that convey the intended meaning.

Cultural Relevance: Ensure that questions, examples, and scenarios used in the instrument resonate with the cultural context of the participants.

Avoid Biases: Be cautious of bias or assumptions that could arise from your cultural perspective. Frame questions neutrally and avoid leading questions.

Sensitive Topics: Approach sensitive topics with care, considering cultural taboos and potential discomfort.

Inclusive Response Options: Provide response options that are relevant and meaningful within the cultural context.

Visual Aids: Use visuals that are culturally relevant and appropriate, taking into account symbols, colors, and representations.

Pilot Testing: Conduct pilot testing within the cultural context to identify any aspects that may need modification.

Flexibility: Be open to adapting the instrument based on participant feedback and cultural insights.

Informed Consent: Clearly explain the research process and goals to participants in a way that aligns with their cultural norms.

9.1.3 Challenges and Considerations

Cultural Diversity: Different cultures have unique communication styles, values, and perceptions. Strive for instruments that are flexible and adaptable to various cultural contexts.

Avoid Stereotyping: Avoid generalizations or stereotypes about a particular culture. Individuals within the same culture may have diverse perspectives.

Subtle Cultural Nuances: Be attentive to subtle nuances in language, gestures, and behaviors that may convey different meanings.

Historical Context: Understand the historical context of the culture you are studying, as it can influence participants' responses.

Ethical Approval: Seek ethical approval from relevant authorities or review boards in the cultural context, especially when working with indigenous communities or minority groups.

By prioritizing cultural sensitivity in instrument design, researchers can enhance the validity and credibility of their data collection efforts across diverse cultural settings. Ensuring that the research process is respectful and relevant to participants' cultural contexts contributes to the overall success and ethical conduct of the study.

9.2 Adapting Tools for Different Demographics

Adapting data collection tools for different demographic groups involves tailoring instruments to suit the unique characteristics, needs, and preferences of specific populations. This process ensures that data collection is effective and relevant, even when working with diverse groups such as age ranges, genders, educational backgrounds, or socio-economic statuses. In this section, we explore the importance of adapting tools for different demographics and provide guidelines for accomplishing this task.

9.2.1 Significance of Adapting Tools for Different Demographics

Accuracy and Validity: Adapting tools to demographic characteristics enhances the accuracy and validity of the collected data, as participants can better relate to questions and response options.

Participant Engagement: Tailored tools are more likely to engage participants, as they feel that the instrument resonates with their personal experiences and perspectives.

Minimized Bias: Adapting tools help minimize bias by avoiding assumptions that may not apply to specific demographic groups.

Data Representativeness: Tools that are inclusive and adaptable ensure that the data collected is representative of the entire target population.

9.2.2 Guidelines for Adapting Tools for Different Demographics

Understand Demographic Characteristics: Begin by understanding the characteristics, needs, preferences, and cultural norms of the specific demographic group.

Language and Literacy: Use language that is appropriate for the literacy level of the target demographic. Ensure that terms are understandable and relatable.

Sensitivity to Gender and Identity: Be inclusive of different gender identities and avoid binary assumptions. Provide options that resonate with participants' gender experiences.

Age-Appropriate Content: Tailor questions to be age-appropriate, considering the cognitive abilities and developmental stages of participants.

Visual Elements: Incorporate visuals that appeal to the demographic group, considering cultural references and preferences.

Diverse Examples: Include examples and scenarios that participants can relate to, reflecting their experiences and context.

Cultural Relevance: Adapt content to align with cultural norms and values, avoiding topics or phrasings that may be culturally insensitive.

Accessibility: Ensure that the data collection process is accessible to individuals with disabilities, providing necessary accommodations.

Engagement Strategies: Use engagement techniques that resonate with the demographic group, such as gamification, storytelling, or interactive elements.

Pilot Testing: Test the adapted tool with representatives from the demographic group to gather feedback and identify potential adjustments.

9.2.3 Challenges and Considerations

Avoid Stereotypes: While adapting tools, avoid perpetuating stereotypes or making assumptions about the demographic group.

Balancing Inclusivity: Strive for inclusivity while ensuring that the adapted tool does not become overwhelming or confusing for participants.

Subgroups within Demographics: Recognize that even within a specific demographic, there can be diversity in experiences, perspectives, and needs.

Ethical Considerations: Address any potential ethical concerns that may arise when adapting tools for sensitive or vulnerable demographic groups.

Researcher Bias: Be aware of potential researcher bias when making adaptations, and seek input from representatives of the demographic group.

Adapting data collection tools for different demographics requires a careful and thoughtful approach. By respecting the unique characteristics and experiences of various groups, researchers can create instruments that facilitate meaningful and accurate data collection while ensuring inclusivity and ethical considerations.

9.3 Cross-Cultural Challenges and Solutions

Conducting data collection across diverse cultural contexts presents unique challenges due to differences in communication styles, norms, and worldviews. Successfully navigating these challenges requires researchers to be culturally aware, adaptable, and respectful. In this section, we delve into the cross-cultural challenges faced during data collection and provide strategies for overcoming them.

9.3.1 Common Cross-Cultural Challenges

Language Barriers: Communication difficulties can arise when participants and researchers do not share a common language.

Misinterpretation: Cultural differences in interpretation can lead to misunderstandings of questions or responses.

Cultural Norms: Norms around disclosure, privacy, and authority can affect participants' willingness to share information openly.

Social Hierarchies: Hierarchical societies may lead to participants providing responses they believe the researcher expects, rather than their true opinions.

Cultural Bias: Unintentional bias in instrument design may not resonate with participants from different cultural backgrounds.

9.3.2 Strategies for Overcoming Cross-Cultural Challenges

Cultural Sensitivity Training: Equip researchers with training in cross-cultural communication and awareness to navigate cultural differences effectively.

Bilingual Assistants: Employ bilingual assistants or interpreters to bridge language gaps and facilitate communication.

Culturally Tailored Instruments: Design instruments that are culturally relevant and resonate with participants' experiences.

Pilot Testing: Pilot test instruments in the cultural context to identify language or interpretation issues and make necessary adjustments.

Cultural Mediation: Collaborate with local experts who can mediate interactions and provide insights into cultural norms.

Flexible Approach: Be flexible and open to adapting the research process to align with cultural practices and expectations.

Build Trust: Invest time in building trust with participants through respectful engagement and open communication.

Respect Hierarchies: Understand social hierarchies and use appropriate communication styles when interacting with participants.

Cultural Empathy: Put yourself in participants' shoes to understand their perspectives and potential challenges in sharing information.

9.3.3 Challenges and Considerations

Ethical Sensitivity: Cultural differences may lead to ethical dilemmas. Always prioritize participants' well-being and rights.

Informed Consent: Ensure that participants fully understand the research process and their rights, even if cultural norms affect their response to consent.

Ethical Review: Seek ethical approval from relevant authorities or review boards in the cultural context to ensure alignment with local regulations.

Non-Verbal Communication: Be aware of non-verbal cues and gestures that may carry different meanings in different cultures.

Cultural Humility: Approach cross-cultural interactions with humility and a willingness to learn from participants.

Conducting cross-cultural data collection requires a delicate balance of cultural awareness, adaptability, and ethical considerations. By taking proactive steps to address challenges and tailor the research process to the cultural context, researchers can ensure that data collection is respectful, effective, and meaningful across diverse cultural settings.

Example 1: Data Collection from Third Gender (Transgender) Population

Collecting data from the transgender population requires a sensitive and culturally competent approach that respects their unique experiences, challenges, and identities. Here, we provide an example of how to adapt data collection methods for this specific demographic.

Understanding the Context:

In many cultures, transgender individuals often face discrimination, stigmatization, and lack of legal recognition. They may have unique gender identities that differ from the binary male-female framework.

Cultural Sensitivity in Instrument Design:

1. Language and Identity:

   Develop survey questions that allow participants to self-identify their gender, using inclusive terms like "transgender," "non-binary," "genderqueer," etc.
   Include pronoun options that go beyond he/she, such as they/them, and allow participants to write in their pronouns.

   Avoid using gendered language in questions or scenarios. For example, instead of "What do you think a man's role in society is?" use "What do you think a person's role in society is?"

2. Sensitive Topics:

   Be respectful when asking about personal experiences related to gender identity, discrimination, or mental health.

   Frame questions about experiences of discrimination in a non-leading way to capture genuine experiences. For example, "Have you ever experienced discrimination based on your gender identity?"

3. Privacy and Confidentiality:

   Assure participants that their responses will be confidential and will not be used against them.

   Use pseudonyms or participant numbers instead of real names to maintain anonymity.

Adapting Data Collection Methods:

1. Interviews:

   Use open-ended questions to allow participants to share their experiences and narratives.

   Approach interviews with empathy and respect, acknowledging the importance of their stories.

2. Surveys:

   Provide an option to skip questions that participants might find uncomfortable or invasive.

   Use a mix of closed-ended and open-ended questions to allow participants to express themselves fully.

3. Focus Group Discussions:

   Facilitate an inclusive environment where participants feel safe to share their thoughts.

   Include icebreaker questions to build rapport and create a comfortable atmosphere.

4. Observations:

   Seek consent before observing transgender individuals in any context, respecting their autonomy and privacy.

Cultural Considerations:

Be aware of diverse transgender experiences across different cultures and regions. Avoid generalizations.

Collaborate with transgender community leaders or organizations to ensure cultural sensitivity and appropriateness.

Ethical Considerations:

Obtain informed consent that clearly outlines the purpose of the research and how the data will be used.

Be prepared to provide emotional support or resources if participants become distressed during the data collection process.

Collecting data from the transgender population requires an inclusive and empathetic approach that respects their identities and experiences. By adapting data collection methods and instruments with cultural sensitivity in mind, researchers can create a safe and respectful environment for transgender individuals to share their stories and perspectives.

Example 2: Data Collection from Ethnic Minority Groups

Collecting data from ethnic minority groups necessitates a culturally sensitive and respectful approach that recognizes their unique cultural, social, and historical contexts. Here, we provide an example of how to adapt data collection methods for this specific demographic.

Understanding the Context:

Ethnic minority groups often possess distinct cultural practices, languages, and historical experiences that influence their perspectives and responses. Respect for their cultural identity and historical context is paramount.

Cultural Sensitivity in Instrument Design:

1. Inclusive Language:

   Incorporate terms and phrases that resonate with the specific ethnic group's cultural identity and experiences.

   Use bilingual or multilingual instruments if participants are more comfortable responding in their native language.

2. Cultural Practices:

   Incorporate questions that explore cultural practices, traditions, or customs that might influence their opinions or experiences.

   Be cautious not to assume that all members of the ethnic minority group share the same cultural traits.

3. Contextual Examples:

   Frame questions using examples or scenarios that are relevant to the lived experiences of the ethnic minority group.

   Ensure that these examples are respectful and avoid cultural stereotypes.

Adapting Data Collection Methods:

1. Interviews:

   Engage in active listening, allowing participants to express their perspectives and experiences in their own words.

   Tailor interview questions to reflect the cultural context of the ethnic minority group.

2. Surveys:

   Ensure that response options are culturally relevant and encompass a diverse range of perspectives.

   Use skip logic to allow participants to answer questions that are most applicable to their experiences.

3. Focus Group Discussions:

   Foster an environment where participants feel comfortable discussing cultural practices, traditions, and perspectives.

Facilitate discussions that allow participants to share unique insights from their cultural backgrounds.

4. Observations:

   Pay attention to cultural cues, gestures, and behaviors during observations, recognizing their cultural significance.

Cultural Considerations:

Collaborate with community leaders or members of the ethnic minority group to ensure cultural authenticity and relevance.

Avoid making assumptions about the experiences or viewpoints of individuals within the ethnic minority group.

Ethical Considerations:

Obtain informed consent that outlines the research purpose and the intended use of the data.

Be mindful of power dynamics, ensuring that participants feel empowered to decline participation without consequences.

Collecting data from ethnic minority groups requires a nuanced approach that acknowledges their cultural diversity and respects their historical context. By adapting data collection methods and instruments with cultural sensitivity in mind, researchers can create a respectful and inclusive space for ethnic minority individuals to share their perspectives and experiences.

Example 3: Data Collection from Sex Workers

Collecting data from sex workers demands a sensitive and nonjudgmental approach that recognizes their unique challenges, vulnerabilities, and experiences. Here, we provide an example of how to adapt data collection methods for this specific demographic.

Understanding the Context:

Sex workers often face stigma, legal challenges, and societal marginalization. It's crucial to approach data collection with empathy and a commitment to respecting their agency and rights.

Cultural Sensitivity in Instrument Design:

1. Nonjudgmental Language:

   Use neutral and nonjudgmental language throughout the instrument to avoid further stigmatization.

   Address the diversity within the sex work community and acknowledge different work settings and experiences.

2. Informed Consent:

   Clearly explain the purpose of the research and how the data will be used, ensuring participants fully understand before providing consent.

   Highlight participants' rights to privacy and confidentiality.

Adapting Data Collection Methods:

1. Interviews:

   Use open-ended questions to allow participants to share their experiences, challenges, and perspectives.

   Approach interviews with empathy and respect for their agency, avoiding judgmental or leading questions.

2. Surveys:

   Prioritize anonymity and confidentiality in surveys, assuring participants that their responses will be kept confidential.

   Allow participants to skip questions that they find uncomfortable or invasive.

3. Building Trust:

   Develop rapport with participants before delving into sensitive questions, ensuring they feel comfortable sharing their experiences.

Cultural Considerations:

Be aware of the diversity within the sex work community, recognizing the different circumstances and challenges individuals may face.

Collaborate with organizations or advocates that work with sex workers to ensure cultural sensitivity and relevance.

Ethical Considerations:

Prioritize the well-being and rights of sex workers throughout the data collection process.

Offer resources and information on support services available to sex workers, in case participants require assistance.

Collecting data from sex workers requires a trauma-informed and compassionate approach that recognizes their autonomy and rights. By adapting data collection methods and instruments with sensitivity to their experiences, researchers can create a safe and respectful space for sex workers to share their perspectives and contribute to meaningful research.

Example 4: Data Collection from Refugee Communities

Collecting data from refugee communities necessitates a culturally sensitive approach that respects their unique experiences, challenges, and contexts. Here, we provide an example of how to adapt data collection methods for this specific demographic.

Understanding the Context:

Refugee communities often have diverse backgrounds, cultural norms, and experiences due to displacement. Approach data collection with empathy, acknowledging their resilience and vulnerabilities.

Cultural Sensitivity in Instrument Design:

1. Trauma-Informed Language:

   Use language that avoids triggering traumatic memories or experiences related to displacement.

Frame questions in a way that is considerate of potential emotional triggers.

2. Informed Consent:

   Explain the research objectives and how the data will be used clearly and transparently.

   Provide information on confidentiality and address any concerns about safety and security.

Adapting Data Collection Methods:

1. Interviews:

   Begin interviews with open-ended questions that allow participants to share their journey, challenges, and aspirations.

   Respect their experiences and show empathy when discussing sensitive topics related to displacement.

2. Surveys:

   Design surveys with simplicity and clarity to ensure participants from different linguistic and educational backgrounds can understand the questions.

   Use visual aids, symbols, or pictures to aid comprehension.

3. Building Trust:

   Build trust by collaborating with community leaders, advocates, or organizations that work closely with the refugee community.

   Create a safe and comfortable environment for participants to share their stories.

Cultural Considerations:

Recognize that refugees come from diverse cultural backgrounds and may have varying levels of acculturation.

Adapt methods to be inclusive of different languages and dialects spoken within the community.

Ethical Considerations:

Be sensitive to participants' potential trauma and ensure they are aware of available support services.

Prioritize their well-being, and avoid re-traumatization by respecting boundaries and emotional readiness.

Collecting data from refugee communities requires a trauma-informed and culturally respectful approach that acknowledges their past experiences and current challenges. By adapting data collection methods and instruments with sensitivity to their unique circumstances, researchers can create a safe and inclusive space for refugees to share their stories and contribute to meaningful research.

CHAPTER 10

Analyzing and Interpreting Collected Data

ABSTRACT
This chapter searches into the critical phase of making sense of gathered information in research. It commences by elucidating the pivotal steps involved in data cleaning and preparation for analysis, emphasizing the significance of pristine data. Subsequently, it delineates the varied approaches to both quantitative and qualitative data analysis, encompassing statistical techniques and qualitative methodologies. The chapter culminates by elucidating the process of drawing well-informed conclusions from the analyzed data, underlining its relevance in aiding evidence-based decision-making.

Keywords: Data cleaning, data preparation, quantitative analysis, qualitative analysis, informed decisions

10.1 Data Cleaning and Preparation for Analysis

Before diving into the analysis phase, it's essential to ensure that the collected data is accurate, complete, and ready for meaningful interpretation. Data cleaning and preparation are critical steps that involve identifying and rectifying errors, inconsistencies, and missing values. This section outlines the importance of data cleaning and provides guidance on how to prepare the collected data for analysis.

10.1.1 Importance of Data Cleaning and Preparation

Accuracy: Clean data reduces the risk of drawing incorrect conclusions due to errors or inconsistencies.

Reliability: Accurate and consistent data ensures that the findings are reliable and trustworthy.

Quality Analysis: High-quality data is crucial for conducting meaningful and insightful analysis.

Effective Interpretation: Well-prepared data allows for accurate interpretation of trends, patterns, and correlations.

10.1.2 Guidelines for Data Cleaning and Preparation

Data Validation: Verify the accuracy and completeness of each data entry against the source.

Missing Data: Identify and address missing data points. Decide whether to impute missing values or exclude incomplete cases.

Outliers: Identify and handle outliers that might skew the analysis. Determine whether outliers should be retained or treated.

Consistency Checks: Ensure consistency across variables and responses. Cross-check for data entry errors.

Standardization: Standardize variables if necessary, like converting dates or categorizing open-ended responses.

Data Coding: Assign appropriate codes to categorical variables for easier analysis.

Labeling: Clearly label variables and provide a codebook detailing the meaning of each variable and its possible values.

Data Transformation: Transform data if required, such as calculating percentages or creating new variables.

Data Entry Errors: Correct any typographical errors that might have occurred during data entry.

Data Security: Maintain data security and privacy throughout the cleaning process.

10.1.3 Software Tools for Data Cleaning

Various software tools can assist in data cleaning and preparation, such as Microsoft Excel, Python (with libraries like Pandas), R, and dedicated data cleaning software.

10.1.4 Challenges and Considerations

Time and Resources: Data cleaning can be time-consuming, especially with large datasets.

Subjectivity: Deciding on data treatment methods can involve subjective judgment, potentially impacting results.

Loss of Information: Aggressive cleaning might lead to loss of valuable information. Balance between cleaning and retaining data.

Documentation: Maintain thorough documentation of the cleaning steps taken, making the analysis reproducible.

Ethical Concerns: Ensure that data cleaning and preparation respect ethical considerations, especially when handling sensitive data.

Effective data cleaning and preparation set the foundation for meaningful analysis and interpretation. By ensuring that the collected data is accurate, consistent, and ready for analysis, researchers can confidently move forward to uncover insights, trends, and patterns that inform the research objectives.

10.2 Approaches to Quantitative and Qualitative Analysis

After data cleaning and preparation, the next step is to analyze the collected data. The approach to analysis depends on whether the data is quantitative (numeric) or qualitative (textual or narrative). This section explores the distinct approaches for both types of data analysis.

10.2.1 Quantitative Analysis

Quantitative Analysis is a structured and systematic approach to processing numerical data, which allows researchers to uncover patterns, relationships, and statistical significance within their dataset. This method employs various techniques, each serving a specific purpose in data analysis:

Descriptive Statistics:

Mean: The average value of a dataset.

Median: The middle value when data is ordered.

Mode: The most frequently occurring value.

Range: The difference between the maximum and minimum values.

Standard Deviation: A measure of data dispersion from the mean.

Inferential Statistics:

T-Tests: Used to compare the means of two groups to determine if there is a significant difference.

Analysis of Variance (ANOVA): Extends the t-test to compare means across more than two groups.

Regression Analysis: Examines the relationship between a dependent variable and one or more independent variables.

Chi-Square Test: Assesses the association between categorical variables.

Correlation and Regression:

Correlation Analysis: Measures the strength and direction of a relationship between two continuous variables.

Linear Regression: Predicts a dependent variable based on the linear relationship with one or more independent variables.

Logistic Regression: Used when the dependent variable is binary, predicting probabilities of one of two possible outcomes.

Hypothesis Testing:

Null Hypothesis (H0): A statement of no effect or no difference.

Alternative Hypothesis (Ha): A statement that contradicts the null hypothesis.

Significance Level (α): The probability of rejecting the null hypothesis when it is true (commonly set at 0.05).

P-Value: The probability of obtaining test results as extreme as the ones observed, assuming the null hypothesis is true.

Data Visualization:

Histograms: Depict the distribution of a single variable.

Scatter Plots: Illustrate relationships between two continuous variables.

Bar Charts: Display categorical data.

Line Charts: Show trends over time.

Box Plots: Visualize data distribution, including outliers.

Heatmaps: Depict the correlation matrix between variables.

Quantitative analysis is particularly valuable when dealing with large datasets and research questions that involve measuring the magnitude and significance of associations. It aids in drawing objective conclusions based on numerical evidence and provides a foundation for making informed decisions in various fields, including science, economics, social research, and more.

10.2.2 Qualitative Analysis

Qualitative Analysis is a comprehensive approach used to decipher textual or narrative data, unveiling profound themes, patterns, and concealed meanings. Several strategies are employed to dissect qualitative data, each serving a specific purpose in the interpretation:

Thematic Analysis:

Objective: To identify recurring themes and patterns within qualitative data.

Process: Researchers systematically code data, grouping similar content into themes or categories.

Outcome: It reveals prevalent themes, providing insight into the dataset's underlying structure.

Content Analysis:

Objective: To scrutinize the content of text, pinpointing keywords and themes for data categorization and interpretation.

Process: Researchers extract keywords and phrases, categorizing content into themes.

Outcome: It enables researchers to dissect textual data efficiently, unveiling content-driven insights.

Narrative Analysis:

Objective: To dissect the structure and content of narratives, uncovering storytelling techniques and the construction of meaning.

Process: Researchers examine narrative elements like plot, characters, and discourse to comprehend storytelling strategies.

Outcome: It yields a profound comprehension of how narratives shape meaning.

Grounded Theory:

Objective: To construct theories grounded in data rather than preconceived notions, permitting novel insights to emerge.

Process: Researchers engage with data iteratively, forming concepts and hypotheses throughout the analysis.

Outcome: It generates theory directly from the data, fostering a deeper understanding of the subject.

Discourse Analysis:

Objective: To dissect language usage, exposing underlying social, cultural, or power dynamics.

Process: Researchers scrutinize language structure, metaphors, and linguistic devices.

Outcome: It unveils how language constructs reality and exposes embedded societal constructs.

Qualitative analysis is fundamental when dealing with rich, textual data sources, such as interviews, focus groups, or written content. It allows researchers to explore complex phenomena, contexts, and subjective experiences, offering a deeper understanding of the qualitative aspects of research questions. This approach is commonly applied in social sciences, humanities, anthropology, and related fields where textual and narrative data are prevalent.

10.2.3 Mixed-Methods Analysis

In certain research scenarios, researchers employ a mixed-methods approach, which amalgamates both quantitative and qualitative analysis methods. This comprehensive technique is employed to attain a deeper understanding of multifaceted phenomena by integrating data from both sources and applying diverse analytical strategies.

Choosing the Right Approach:

The selection of an appropriate analysis approach hinges on various factors:

Research Objectives: The primary research objectives and questions guide the choice of analysis methods.

Data Type: The nature of the data, whether quantitative, qualitative, or both, plays a pivotal role.

Researcher Expertise: Researchers must consider their proficiency in quantitative and qualitative analysis techniques.

Available Resources: The availability of tools, software, and human resources affects the chosen approach.

Desired Depth: Researchers must decide the level of depth and granularity required for analysis.

Challenges and Considerations:

Several considerations and challenges accompany mixed-methods analysis:

Data Interpretation: Adequate interpretation of both quantitative and qualitative data is essential to avoid erroneous conclusions.

Subjectivity: Qualitative analysis inherently involves subjective interpretation, potentially introducing bias.

Software Tools: Researchers can utilize specialized software tools such as SPSS, R, Python for quantitative analysis and NVivo or ATLAS.ti for qualitative analysis.

Time and Resources: The combined approach can be resource-intensive, necessitating careful planning.

Validity and Reliability: Upholding validity and reliability across both types of data and analyzes is crucial to maintain research integrity.

Researchers often opt for mixed-methods analysis when tackling multifaceted research questions or when a more comprehensive exploration of a phenomenon is required. This approach allows them to leverage the strengths of both quantitative and qualitative data, providing a well-rounded perspective on the research topic.

Effective analysis provides insights that answer research questions and contribute to knowledge. Whether using quantitative, qualitative, or mixed-methods approaches, careful execution and interpretation are essential to deriving meaningful conclusions from the collected data.

10.3 Drawing Conclusions and Making Informed Decisions

Drawing conclusions from analyzed data is a critical step that guides informed decision-making, contributes to research outcomes, and informs future actions. This section explores how to effectively draw conclusions from data analysis and use those conclusions to make informed decisions.

10.3.1 Interpreting Findings

Interpreting the findings of your data analysis is a crucial phase that bridges the gap between raw data and actionable insights. It involves making sense of the collected information, discerning meaningful patterns, and deriving conclusions that align with your research objectives. Effectively interpreting findings not only validates your research efforts but also empowers you to contribute to knowledge, make informed decisions, and drive positive change. Here are key aspects to consider when interpreting your findings:

Data Alignment: Ensuring Direct Alignment with Research Objectives

The foundation of meaningful interpretation lies in ensuring that your conclusions are firmly rooted in the data you collected and analyzed. Always reference back to your research objectives and questions to verify that your findings directly address them. This alignment enhances the credibility of your conclusions, reinforcing that your research objectives were met and your analysis was on target.

Patterns and Trends: Identifying Significant Insights

In the vast sea of data, it's essential to identify and highlight significant patterns, trends, correlations, or themes that emerged during analysis. These insights offer a lens through which you can understand the underlying dynamics of your research phenomenon. Whether it's a spike in quantitative data indicating a particular trend or a recurring qualitative theme that echoes throughout interviews, these patterns guide your understanding and help you formulate informed conclusions.

Statistical Significance: Quantitative Data's Verdict

In quantitative research, statistical significance plays a pivotal role in drawing conclusions about relationships between variables. When analyzing relationships or differences between groups, statistical tests help determine if observed patterns are unlikely to have occurred by chance. Acknowledging statistical significance adds a layer of objectivity to your conclusions, demonstrating that your findings are robust and not mere coincidences.

Qualitative Insights: Unearthing Richness from Narratives

For qualitative data, the interpretation process delves into the richness of narratives, stories, and experiences shared by participants. Your conclusions should reflect the essence of these qualitative insights, capturing the nuance and depth that emerges from open-ended responses, interviews, or focus group discussions. Identifying key themes, central narratives, and poignant quotes showcases the unique perspectives that participants bring to your research.

Visual Representation: Enhancing Clarity

Visual aids, such as graphs, charts, or diagrams, can be powerful tools to support your interpretation. They condense complex information into easily digestible visuals, allowing your audience to grasp patterns and relationships at a glance. Visual representation not only adds clarity to your conclusions but also reinforces the validity of your findings.

Balancing Complexity and Simplicity: Effective Communication

While your analysis might be intricate, your conclusions should be communicated in a clear, concise, and accessible manner. Strive to strike a balance between providing the necessary depth of insight and avoiding overwhelming your audience with technical jargon. Effective communication ensures that your conclusions resonate with a diverse range of stakeholders, from fellow researchers to policymakers and the general public.

Interpreting findings transforms data into knowledge and insights. By staying aligned with research objectives, identifying meaningful patterns, considering statistical significance, valuing qualitative richness, and communicating effectively, you unlock the potential of your data to contribute to understanding, decision-making, and positive impact.

10.3.2 Making Informed Decisions

Concluding data analysis is just the beginning. The true impact of your research lies in how well those conclusions inform decisions and actions, and contribute to the broader knowledge landscape. Here's a closer look at key considerations when using your findings to make informed decisions:

Relevance to Objectives: Aligning with Research Intent

As you make decisions based on your conclusions, ensure that they directly correspond to the research questions or objectives you established at the outset. This alignment reinforces the purposefulness of your research and ensures that your conclusions remain focused and relevant.

Practical Implications: Connecting to Real-World Contexts

Translating research findings into practical implications involves understanding how your conclusions apply in real-world contexts. Consider how your insights can be practically applied or adapted to solve problems, address challenges, or enhance existing practices.

Policy and Practice: Influencing Decision-Making

The impact of research often extends to policy, practice, and decision-making in relevant fields. Examine how your conclusions can contribute to informed decision-making processes. Policymakers, practitioners, and stakeholders can use your insights to shape strategies, allocate resources, and enact positive change.

Recommendations: Guiding Future Actions

If your research uncovers actionable insights, consider providing recommendations based on your conclusions. These recommendations can guide future actions, interventions, or projects. Whether it's suggesting best practices, proposing changes to existing procedures, or highlighting areas that warrant further exploration, your recommendations add value beyond your research.

Balancing Feasibility and Aspiration:

When making decisions based on your conclusions, consider the balance between what's feasible in the immediate term and what could be aspirational in the long run. Practical implications and policy considerations might involve incremental steps, while recommendations could offer visionary paths for advancement.

Communication: Bridging the Gap

Effectively communicating your conclusions is essential to ensure that they influence the right stakeholders. Tailor your communication to resonate with policymakers, practitioners, or the general public, using language and visuals that suit their level of expertise.

Ethical Reflection: Ethical Implications and Responsibility

Pause to reflect on the ethical implications of your conclusions. Consider potential risks, unintended consequences, and the ethical responsibility that comes with disseminating findings that could impact individuals, communities, or society at large.

Continued Learning: Iteration and Exploration

Remember that conclusions are not endpoints, but rather stepping stones for further exploration and learning. Use your conclusions to inform future research, broaden your understanding, and contribute to ongoing discussions within your field.

Making informed decisions based on research conclusions is a powerful way to drive change and advance knowledge. By ensuring alignment with objectives, considering practical implications, influencing policy and practice, offering recommendations, and communicating effectively, you transform your research into a catalyst for positive impact.

10.3.3 Transparency and Limitations

Maintaining transparency and acknowledging limitations are integral aspects of responsible research practice. They ensure that your research process and findings are presented honestly and ethically, allowing stakeholders to critically evaluate your work and its implications. Here's a closer look at the significance of transparency and addressing limitations:

Transparency: Clear Articulation of Analysis Steps

Transparency involves clearly and comprehensively detailing the steps you undertook during data analysis. It's about providing a roadmap for others to follow and understand how you arrived at your conclusions. Key elements of transparency include:

- Methods and Assumptions: Articulate the specific methods, tools, and software used for analysis. Explain any assumptions you made during the process.
- Data Transformation: Describe any transformations, cleaning procedures, or adjustments made to the raw data before analysis.
- Statistical Tests: If working with quantitative data, explain the statistical tests used and the rationale behind their selection.
- Qualitative Analysis: For qualitative data, outline the approach to coding, thematic analysis, and the steps taken to ensure rigor.
- Interpretations: Clearly explain how you interpreted the findings, connecting them back to the research objectives.
- Visual Aids: If you used visual aids, describe each chart, graph, or diagram along with its purpose.

Limitations: Recognizing Research Boundaries

Every research study has limitations that impact the generalizability and robustness of findings. Addressing these limitations demonstrates your awareness of potential weaknesses in your study and your commitment to presenting a balanced perspective. Common limitations include:

- Sample Size: Acknowledge if your sample size was small, which might affect the generalizability of findings.
- Data Quality: Address any data quality issues, such as missing data or inaccuracies, and discuss how you mitigated their impact.
- Selection Bias: If certain groups were overrepresented or underrepresented in your sample, acknowledge the potential for selection bias.
- Methodological Constraints: Discuss any limitations in your chosen methods, such as limitations in survey questions or the inability to control extraneous variables.

- External Validity: Address the extent to which your findings can be applied beyond the specific context of your study.
- Ethical Considerations: Reflect on any ethical dilemmas faced during the research and how they might have influenced the findings.

The Value of Transparency and Limitations:

Transparency and acknowledging limitations enhance the credibility of your research and foster an environment of intellectual honesty. They enable fellow researchers to replicate or build upon your work and encourage constructive critique. By presenting both the strengths and weaknesses of your study, you contribute to the advancement of knowledge by promoting robust, nuanced discussions within your field.

Ethical Responsibility:

Transparency and limitations align with the ethical responsibility to present research honestly, respecting the potential impact of your findings on individuals, communities, and society. Openly discussing limitations also helps prevent unwarranted conclusions or actions based on incomplete information.

Continuous Improvement: Learning and Progression

Transparency and addressing limitations also contribute to your growth as a researcher. Reflecting on your study's shortcomings encourages you to refine your research methodologies, anticipate challenges, and design more rigorous studies in the future.

Transparency and addressing limitations are cornerstones of ethical and responsible research practice. By clearly articulating your analysis steps, assumptions, and interpretations, as well as recognizing the boundaries of your study, you foster accountability and integrity in your research process and findings.

10.3.4 Communication and Reporting

Effectively communicating your research findings is pivotal to ensure that your hard work has a meaningful impact. It's the bridge that connects your insights to the stakeholders who can benefit from them. Here's how to optimize your communication and reporting strategies:

Audience-Centric Approach: Relevance to Different Stakeholders: Tailor your conclusions to resonate with your target audience. Consider who will be reading or engaging with your report—whether policymakers, practitioners, fellow researchers, or the general public. Adapting your language, tone, and depth of explanation ensures that your message is received and understood by your intended audience.

Clear Communication: Making Complexity Understandable: Research findings can be intricate, but your communication doesn't have to be. Present your conclusions in a clear, concise, and accessible manner. Break down complex concepts, avoid jargon, and use plain language to convey your insights without overwhelming your audience.

Visuals and Graphics: Enhancing Clarity: Visual aids like graphs, charts, and diagrams can be invaluable tools in simplifying complex data. Visuals provide an instant understanding of trends, relationships, and patterns, making your conclusions more engaging and comprehensible. Ensure that each visual is labeled, easily interpretable, and relevant to your narrative.

Supporting Evidence: Reinforcing Conclusions: Back your conclusions with concrete evidence. Use relevant data points, statistical summaries, charts, graphs, and direct quotes from qualitative analysis to substantiate your claims. This not only lends credibility to your conclusions but also offers a more comprehensive picture of the insights you're presenting.

Contextualization: Bridging the Gap: Connect your conclusions to the larger context. Explain how your findings fit into the broader field of research, address existing gaps, or contribute to ongoing conversations. Providing this context helps your audience appreciate the significance of your conclusions.

Engagement: Inviting Interaction: Encourage engagement with your findings. Consider using storytelling techniques to bring your research to life and evoke empathy or curiosity. Pose thought-provoking questions, scenarios, or challenges that encourage your audience to think critically about the implications of your research.

Ethical Communication: Transparency and Honesty: In your reporting, maintain the same transparency and honesty that you applied throughout your research process. If there are limitations, uncertainties, or areas where more research is needed, acknowledge them. Ethical communication builds trust and showcases your commitment to responsible research practices.

Feedback and Iteration: Learning from Interaction: As you communicate your findings, be open to feedback. Constructive critique can help you refine your conclusions, better address potential misunderstandings, and identify areas for improvement in your research or communication strategy.

Impact and Legacy: The Power of Effective Communication: Effective communication amplifies the impact of your research. It allows your conclusions to reach the right people, spark conversations, influence decision-making, and contribute to positive change. Your research has the potential to shape policies, inform practices, and inspire future studies, making clear and engaging communication an essential component of your research journey.

By tailoring your communication to your audience, embracing clear language and visuals, providing supporting evidence, contextualizing your findings, promoting engagement, and adhering to ethical communication principles, you maximize the reach and influence of your research conclusions.

10.3.5 Ethical Considerations

Ethical considerations are a cornerstone of responsible research practice, guiding how you handle sensitive information, vulnerable populations, and personal data. As you draw conclusions and communicate your findings, keep these ethical considerations at the forefront:

Ethical Implications: Navigating Sensitive Topics: When concluding, particularly from research involving sensitive topics or vulnerable populations, be acutely aware of the potential ethical implications of your findings. Consider how your conclusions might impact individuals, communities, or societal perceptions. Strive to present your findings responsibly, avoiding sensationalism or exploitation.

Balancing Transparency and Privacy: While transparency is crucial, it's equally important to prioritize the privacy and confidentiality of your participants. When sharing your conclusions, ensure that you do not reveal any personally identifiable information. Use pseudonyms or aggregate data to maintain the anonymity of individuals, especially when discussing case studies or qualitative narratives.

Informed Consent and Respect: Reflect on whether your conclusions align with the informed consent participants provided. If your findings involve sensitive or unexpected outcomes, consider whether these were adequately communicated to participants during the informed consent process. Respect their right to be informed about how their data is being used and the potential consequences.

Avoiding Harm: Do No Harm Principle: Apply the principle of "not harm" to your conclusions. Reflect on whether your findings have the potential to cause harm, stigmatize individuals, or perpetuate stereotypes. If there's a risk of harm, take measures to mitigate it, such as framing your conclusions in a way that avoids negative consequences.

Beneficence and Justice: Benefitting Participants: Consider how your conclusions can benefit the participants or communities involved in your research. If possible, highlight positive outcomes, potential improvements, or areas where your findings could lead to positive change. Strive to ensure that your research contributes positively to the well-being of those who participated.

Ethics Review and Guidelines: Adhering to Standards: Throughout your research journey, adhere to ethical guidelines and protocols set by your institution or relevant bodies. If you engaged in an ethics review process before conducting your research, ensure that your conclusions and communication align with the approved ethical framework.

Cultural Sensitivity: Context Matters: Take into account the cultural context in which your research was conducted. Consider how your conclusions might resonate or be interpreted differently in various cultural settings. Be sensitive to cultural nuances, avoid cultural appropriation, and respect the perspectives of the communities you engage with.

Continuous Reflection: Ethical Iteration: Ethical considerations are not a one-time checkbox; they require ongoing reflection and adaptation. Continuously assess how your conclusions impact various stakeholders, and be prepared to make ethical adjustments if new insights or perspectives emerge.

Legacy of Responsibility: Ethical Impact: By adhering to ethical considerations in your conclusions and communication, you uphold the integrity of your research and contribute to a legacy of responsible research practice. Ethical decisions amplify the positive impact of your research and set an example for future researchers to follow.

Incorporating ethical implications into your conclusions involves a delicate balance of transparency, privacy, respect, and beneficence. By considering the potential ethical consequences of your conclusions and prioritizing the well-being of participants and communities, you uphold the ethical integrity of your research journey.

10.3.6 Reflection and Iteration

The journey of research doesn't conclude with your findings and conclusions—it continues through reflection and iteration. Embracing these practices enriches your research experience, refines your skills, and contributes to the ongoing growth of knowledge. Here's why reflection and iteration matter:

Reflective Practice: Learning from the Journey: Engage in reflective practice to deeply understand your research journey. Reflect on the choices you made during data collection, the nuances you discovered during analysis, and the insights you derived from your conclusions. Consider the challenges you encountered, the ethical dilemmas you navigated, and the moments of discovery that fueled your curiosity.

Continuous Improvement: Refining Research Skills: Reflection serves as a foundation for continuous improvement. By critically evaluating your research process, you identify areas where you excelled and aspects that could be refined. Use these insights to enhance your research skills, adopt new methodologies, and develop strategies to address challenges more effectively in future projects.

Iterative Process: Evolution of Conclusions: Conclusions are not set in stone; they are part of an iterative process that evolves. Recognize that as you continue your research journey or as new studies emerge, your conclusions might need to be reevaluated or expanded. New data, updated methodologies, or novel insights can reshape your understanding and lead to deeper, more nuanced conclusions.

Staying Current: Embracing New Insights: Research is dynamic, with new information constantly emerging. Keep yourself informed about the latest developments in your field to ensure that your conclusions remain relevant and aligned with current knowledge. This commitment to staying current enriches the impact of your research in academic and practical spheres.

Humility and Openness: Embracing Uncertainty: Embrace humility and acknowledge that research conclusions often exist within a realm of uncertainty. Be open to the idea that your conclusions might evolve as you encounter different perspectives, receive feedback, or explore related studies. This openness fosters a spirit of intellectual curiosity and growth.

Contributing to Knowledge: Advancing Discourse: The iterative process of reflection and refinement contributes to the collective knowledge base. As you refine your research skills and revisit your conclusions, you actively participate in advancing the discourse within your field. Your willingness to learn from your work and the work of others contributes to the continuous progression of understanding.

Personal and Professional Growth: Both reflection and iteration foster personal and professional growth. Reflective practice nurtures self-awareness and a deeper understanding of research nuances. Iterative thinking promotes adaptability and resilience in the face of new information or challenges.

A Cycle of Learning: The journey of research is a cycle of learning, discovery, and improvement. By engaging in reflective practice, refining your skills, embracing iteration, and contributing to the evolution of conclusions, you weave a rich tapestry of intellectual engagement and progress.

10.3.7 Challenges and Considerations

The research journey is marked by challenges that test your skills and shape your growth. Confronting these challenges with resilience and strategic thinking enhances the quality and impact of your conclusions. Here are some key challenges and considerations to navigate:

Ambiguity: Navigating Uncertain Terrain: Research is not immune to ambiguity; findings may sometimes be inconclusive or subject to multiple interpretations. When encountering ambiguity, embrace transparency. Clearly communicate where uncertainties lie, what limitations impact your conclusions, and why certain questions remain unanswered. This transparent approach cultivates credibility and prevents misinterpretation.

Complexity: Simplifying without Dilution: Communicating complex conclusions in a clear and understandable manner is a formidable challenge. Strive to strike a balance between maintaining accuracy and ensuring accessibility. Break down intricate concepts, utilize visuals, and use relatable examples to distill complexity without diluting the essence of your findings.

Balancing Findings: Presenting the Whole Picture: Research often uncovers a spectrum of findings, both positive and negative. Strive for a balanced presentation that includes both the strengths and limitations of your conclusions. Avoid cherry-picking data or focusing solely on favorable outcomes. This holistic approach reflects the integrity of your research and enables a more nuanced understanding.

Applying Findings: From Theory to Action: Conclusions hold value when they can be applied in practical contexts. Ensure that your findings are actionable—capable of guiding decisions, informing policies, or driving change. Consider how your conclusions align with real-world scenarios, and articulate specific steps or strategies that stakeholders can take to apply your insights effectively.

Ethical Considerations: Addressing Ethical Dilemmas: Ethical dilemmas may arise during your research journey that impact your conclusions. Be prepared to navigate these challenges with sensitivity and responsibility. Reflect on how your conclusions may influence individuals, communities, or societal perceptions, and consider ways to mitigate any potential negative consequences.

Cultural Context: Sensitivity to Diverse Perspectives: Conclusions may be interpreted differently across cultural contexts. Be mindful of the potential variations in interpretation and consider diverse perspectives when framing your findings. Ensure that your conclusions do not inadvertently perpetuate cultural insensitivity or biases.

Feedback and Adaptation: A Continuous Process: Challenges are opportunities for growth. Embrace feedback from peers, mentors, or stakeholders to refine your conclusions. Be open to adapting your approach based on constructive critique and new insights. This flexibility enhances the robustness of your conclusions and reflects your commitment to continuous improvement.

Ethical and Responsible Approach: Navigating challenges with an ethical and responsible approach is a hallmark of integrity in research. As you address ambiguity, communicate complexity, balance findings, apply insights, and consider cultural contexts, you uphold the ethical foundation of your work and contribute to the advancement of knowledge.

Embracing these challenges and considerations not only refines your research skills but also fortifies the impact of your conclusions. By demonstrating transparency, empathy, and adaptability, you exemplify the qualities of a thoughtful and responsible researcher.

Drawing conclusions and making informed decisions based on data analysis is the culmination of the research process. By aligning conclusions with research objectives, considering practical implications, and communicating findings effectively, researchers can contribute to knowledge, influence decision-making, and drive positive change in various domains.

CHAPTER 11

Enhancing Data Collection Skills

ABSTRACT
This chapter focuses on the journey of skill development and expertise cultivation in the realm of data collection. It emphasizes the importance of continual improvement of data collection techniques, learning from hands-on experiences, and the pursuit of building expertise in the field. This chapter serves as a guide for researchers to continually refine and advance their data collection skills.

Keywords: Skill development, data collection techniques, hands-on learning, expertise building, continuous improvement

11.1 Continual Improvement of Data Collection Techniques

Data collection is an art that evolves with experience and dedication. Embrace the idea that your skills are not static; they can be refined over time. Here's how you can foster continual improvement in your data collection techniques:

Reflective Practice: Learning from Experience: Regularly reflect on your data collection experiences. Consider what worked well, what challenges you encountered, and how you could enhance your approach. Reflective practice helps you identify areas for growth and refine your strategies.

Feedback Loop: Seeking Constructive Critique: Invite feedback from mentors, peers, or collaborators who have experience in data collection. Their insights can offer fresh perspectives, uncover blind spots, and introduce innovative techniques. Constructive critique accelerates your learning curve and encourages you to explore new avenues.

Learning from Mistakes: Embracing Challenges: View mistakes as valuable learning opportunities. When faced with challenges or setbacks in data collection, analyze what went wrong and why. This reflective analysis equips you to navigate similar situations more effectively in the future.

Staying Informed: Keeping Abreast of Developments: Stay updated with the latest advancements in data collection methodologies. Attend workshops, seminars, and webinars, and engage with research literature to understand emerging trends and best practices. Incorporating new insights into your approach keeps your techniques fresh and relevant.

Experimentation: Trying New Approaches: Don't shy away from experimenting with different data collection techniques. Trying new methods exposes you to diverse approaches and helps you discover what works best in specific contexts. This willingness to explore fosters innovation in your skill set.

Iterative Refinement: Building on Past Experience: Each data collection endeavor builds on the lessons of the previous one. Incorporate feedback, adjust your techniques based on past experiences, and seek opportunities to enhance your approach. Iterative refinement is a hallmark of a skilled data collector.

Technology Integration: Harnessing Digital Tools: Stay abreast of technology that can enhance data collection. Digital tools, mobile apps, and online platforms can streamline processes, improve accuracy, and provide real-time insights. Integrating technology effectively can boost the efficiency of your data collection efforts.

Peer Learning: Sharing and Collaborating: Engage in peer learning by sharing your experiences and collaborating with fellow researchers. Discussions, workshops, and joint projects can expose you to diverse perspectives and deepen your understanding of various data collection techniques.

Personal Growth: Mastery and Confidence: Continual improvement in data collection techniques not only sharpens your professional skills but also boosts your confidence. As you master various approaches, you become a more versatile and adaptable researcher, ready to tackle a wide array of data collection challenges.

Embracing a growth mindset in data collection is a testament to your commitment to excellence. By reflecting on experiences, seeking feedback, experimenting with new approaches, and staying informed, you continually elevate your skills and contribute to the advancement of knowledge in your field.

11.2 Learning from Data Collection Experiences

Every data collection endeavor is a unique opportunity for growth and learning. Extracting insights from your experiences enriches your skills and informs future endeavors. Here's how to effectively learn from your data collection experiences:

Reflective Analysis: Debriefing Your Experience: After each data collection effort, engage in a reflective debriefing. Consider what went well, the challenges you encountered, and the unexpected insights you gained. Reflective analysis helps you identify patterns, strengths, and areas that require improvement.

Identifying Best Practices: What Worked Well: Highlight the aspects of your data collection that proved successful. Did you use effective rapport-building techniques during interviews? Did your observation strategy yield valuable insights? Identifying these best practices allows you to replicate them in future projects.

Challenges and Lessons: Embracing Setbacks: Don't shy away from examining the challenges you faced. What hurdles did you encounter during data collection? How did you address them? Assessing challenges equips you with strategies to mitigate similar issues and empowers you to handle unexpected situations.

Adjustments and Iterations: Tweaking Your Approach: Based on your experiences, consider adjustments or iterations you can make to your data collection techniques. Did you encounter difficulties with certain questions during interviews? Were there logistical obstacles during observations? Use these insights to refine your approach.

Feedback Incorporation: Integrating Constructive Input: If you received feedback from peers, mentors, or participants, evaluate its relevance. Constructive feedback can shed light on blind spots, suggest improvements, and guide you toward a more effective data collection approach.

Evolving Strategies: From Project to Project: Remember that no two data collection projects are identical. Each experience adds to your toolkit of strategies. Adapt your approach based on the context, objectives, and participants of each project to ensure a tailored and effective data collection process.

Documenting Insights: Building a Repository: Maintain a record of your reflections, insights, and lessons learned from each data collection experience. Over time, this repository becomes a valuable resource that you can revisit to inform new projects and avoid repeating past challenges.

Sharing and Learning: Peer Collaboration: Engage in discussions with peers or colleagues who have conducted similar research. Sharing your experiences and hearing about theirs can lead to valuable insights, innovative solutions, and a deeper understanding of effective data collection techniques.

Personal Growth: A Journey of Mastery: Learning from data collection experiences is a journey of personal growth. It builds your expertise, hones your intuition, and equips you with a toolkit of adaptable techniques. As you navigate varied situations, you become a more confident and skilled data collector.

Innovation and Adaptability: Learning from your data collection experiences bolsters your innovation and adaptability. It enables you to anticipate challenges, devise solutions, and fine-tune your approach. Each experience contributes to your evolution as a researcher and enhances the impact of your future data collection efforts.

11.3 Building Expertise in the Field

Becoming an expert in data collection requires dedication, continuous learning, and a commitment to mastering the nuances of research methodology. Here's how to build and showcase your expertise in the field:

Engage with Research Literature: Stay Informed: Regularly immerse yourself in research literature related to data collection methodologies. Explore academic papers, journals, and books to understand the latest advancements, best practices, and emerging trends in the field.

Attend Workshops and Training: Continuous Learning: Participate in workshops, training programs, and webinars that focus on data collection techniques. These opportunities expose you to new methodologies, provide hands-on experience, and allow you to learn from seasoned experts.

Networking and Collaboration: Learn from Others: Connect with researchers, practitioners, and experts in the field. Networking enables you to exchange ideas, share experiences, and learn from others' insights and challenges. Collaborative projects offer opportunities to deepen your expertise through shared learning.

Mentorship: Seek Guidance from Experts: Find mentors who have extensive experience in data collection. Their guidance, advice, and insights can accelerate your learning curve and provide valuable perspectives on refining your techniques.

Contribute to Research: Apply Your Expertise: Apply your expertise in practical research projects. Contributing to research endeavors allows you to showcase your skills, learn from real-world scenarios, and make meaningful contributions to your field.

Teaching and Workshops: Sharing Knowledge: Consider teaching workshops or sharing your expertise with others. Imparting your knowledge to students, colleagues, or aspiring researchers not only strengthens your understanding but also fosters a culture of continuous learning.

Innovate and Experiment: Push Boundaries: Don't hesitate to innovate and experiment with data collection techniques. Push the boundaries of traditional methods, explore hybrid approaches, and integrate technology to advance the field and demonstrate your innovative thinking.

Conduct Reflective Practice: Self-Evaluation: Regularly assess your growth as a data collection expert. Reflect on your journey, identify areas where you've grown, and set goals for continuous improvement. Self-evaluation ensures that you remain focused on building your expertise.

Presentations and Publications: Share Your Insights: Present your research and insights at conferences, seminars, and symposiums. Publishing papers or articles in reputable journals showcases your expertise to a wider audience and establishes you as a recognized contributor in the field.

Peer Review and Evaluation: Expert Assessment: Engage in peer review activities for research papers or data collection methodologies. This not only contributes to the academic community but also demonstrates your expertise in critically evaluating research approaches.

Ethical and Responsible Practice: Integrity Matters: Above all, uphold ethical and responsible research practices. Demonstrating integrity in your research process and conclusions solidifies your reputation as an expert and a trustworthy contributor to the field.

Cultivating Mastery: A Lifelong Journey: Building expertise in data collection is a journey of mastery that extends beyond specific projects. Embrace the continuous learning, networking, and practical application of skills to cultivate expertise that contributes meaningfully to research, knowledge, and innovation.

CHAPTER 12

Conclusion

ABSTRACT
In the concluding chapter, we summarize the key concepts and takeaways from the preceding chapters. This recap serves as a concise reference for researchers. Additionally, we empower readers with a final message, encouraging them to master the art of data collection—a skill that holds the power to shape research, inform decisions, and drive positive change.

Keywords: Key concepts, takeaways, empowerment, data collection mastery

12.1 Recap of Key Concepts and Takeaways

The culmination of your journey through this guide marks not an end, but a beginning—an initiation into the world of effective data collection. As you embark on your research endeavors, remember that the art of data collection is a dynamic process that requires dedication, skill, and ethical responsibility. Let's recap the key concepts and takeaways that you've encountered:

Purpose of Data Collection: Data collection is the backbone of research, enabling evidence-based decision-making, hypothesis validation, and knowledge expansion.

Role of Tools and Instruments: Tools and instruments serve as conduits for gathering data, encompassing methods from interviews to observations, each with its unique merits and challenges.

Data Collection Process: The data collection process involves systematic steps from planning to analysis, necessitating meticulous attention to detail and ethical considerations.

Selecting Tools and Techniques: Choosing the right tools and techniques requires aligning them with research objectives, understanding their strengths and limitations, and adapting them to diverse contexts.

Analysis and Interpretation: Drawing meaningful conclusions involves aligning findings with objectives, identifying patterns, validating through statistics, and presenting insights for various audiences.

Ethical Considerations: Ethical principles underpin the entire data collection journey, emphasizing privacy, consent, cultural sensitivity, and accountability.

Reflection and Iteration: Reflective practice and iterative refinement enhance data collection skills, fostering growth and adaptability.

Expertise Building: Expertise is cultivated through continuous learning, engagement with research literature, collaboration, mentorship, and a commitment to ethical research practice.

In this journey, you've explored the intricacies of data collection methods, learned to navigate challenges, and embraced the responsibility of ethical research. Armed with these insights, you're equipped to embark on your research projects with confidence and competence. Remember that each research endeavor is an opportunity to learn, adapt, and contribute to the ever-evolving landscape of knowledge.

As you venture forth, may your pursuit of effective data collection be marked by curiosity, integrity, and the relentless pursuit of truth. Your contributions to the realm of research hold the potential to shape perspectives, inform decisions, and drive positive change. Embrace the art of data collection as both a science and an art—an art that transforms data points into narratives, and narratives into insights that illuminate the world around us.

Here's to your journey as a skilled data collector, a thoughtful researcher, and a catalyst for progress.

12.2 Empowering Readers to Master Data Collection

As you conclude this guide, you hold the keys to a realm of transformative possibilities. The journey you've undertaken to master data collection is an investment in your growth as a researcher, a thinker, and an agent of change. Here's how this guide empowers you to navigate the intricacies of data collection:

Knowledge and Understanding: You've gained a comprehensive understanding of data collection's significance, its role in research, and the diversity of tools and methods at your disposal. This knowledge lays the foundation for confident decision-making in your research journey.

Methodological Insight: You've delved into the intricacies of various data collection methods—from interviews to observations, focus group discussions to discourse analysis. This insight equips you to select and adapt methods that align with your research objectives and context.

Practical Guidance: The guide has provided you with practical guidelines, checklists, and samples for creating effective data collection instruments. These resources serve as valuable templates that streamline your planning and execution processes.

Ethical Compass: Ethical considerations are woven throughout the guide, highlighting the importance of privacy, consent, cultural sensitivity, and transparency. This ethical compass ensures that your research upholds the highest standards of integrity.

Problem Solving: Challenges and considerations in data collection have been addressed, enabling you to approach ambiguities, complexities, and limitations with resilience and strategic thinking.

Reflective Practice: The guide emphasizes the value of reflective practice and iterative refinement. These principles encourage you to learn from experiences, adjust your techniques, and continuously enhance your skills.

Expertise Cultivation: By exploring how to build expertise and engage with the research community, you're empowered to evolve into a proficient data collector, adept at driving impactful research.

Holistic Understanding: The guide goes beyond techniques, encompassing topics like technology integration, cultural sensitivity, and responsible communication. This holistic understanding equips you to navigate diverse research contexts with finesse.

Confidence and Innovation: Armed with knowledge, practical insights, and ethical principles, you're poised to embark on research projects with confidence and innovation. Your ability to adapt methods, apply findings, and contribute to the research discourse sets you on a trajectory of meaningful impact.

Remember that mastery is a journey, not a destination. Each project is an opportunity to hone your skills, refine your techniques, and deepen your insights. As you venture into the world of research, embrace the challenges, seek knowledge, and stay curious. Let your pursuit of effective data collection be marked by passion, integrity, and the pursuit of understanding.

Your journey as a skilled data collector is not only about gathering information; it's about transforming that information into insights that ignite change, elevate understanding, and contribute to the betterment of society. You are now armed with the tools to wield data collection as a force for positive transformation. The future of research lies in your capable hands—forge ahead and shape it with purpose.

Bibliography

Abowitz, D. A., & Toole, T. M. (2010). Mixed method research: Fundamental issues of design, validity, and reliability in construction research. *Journal of Construction Engineering and Management, 136*(1), 108–116.

Abusaleh, K., & Anwar, A. B. (2022). Research: Meaning and purpose. In M. R. Islam, N. A. Khan, & R. Baikady (Eds), *Principles of social research methodology* (pp. 15–28). Singapore: Springer Nature.

Akhtar, S. (2022). Key informants' interview. In M. R. Islam, N. A. Khan, & R. Baikady (Eds), *Principles of social research methodology* (pp. 389–403). Singapore: Springer Nature.

Azam, M. G. (2022). In-depth case interview. In M. R. Islam, N. A. Khan, & R. Baikady (Eds), *Principles of social research methodology* (pp. 347–364). Singapore: Springer Nature.

Bhattacherjee, A. (2012). *Social science research: Principles, methods, and practices.* <http://repository.out.ac.tz/504/1/Social_Science_Research-_Principles_Methods_and_Practices.pdf>

Biemer, P. P. (2010). Total survey error: Design, implementation, and evaluation. *Public Opinion Quarterly, 74*(5), 817–848.

Biesta, G. (2007). Why "what works" won't work: Evidence-based practice and the democratic deficit in educational research. *Educational Theory, 57*(1), 1–22.

Blommaert, J. (2005). *Key topics in sociolinguistics.* Cambridge: Reino Unido.

Bondarouk, T., & Ruël, H. J. M. (2004). Discourse analysis: Making complex methodology simple. In T. Leino, T. Saarinen, & S. Klein (Eds), *Proceedings of the 12th European Conference on Information Systems (ECIS)*, June 14–16, Turku, Finland. <https://research.utwente.nl/en/publications/discourse-analysis-making-complex-methodology-simple>

Braumoeller, B. F. (2004). Hypothesis testing and multiplicative interaction terms. *International Organization, 58*(4), 807–820.

Bryman, A. (2016). *Social research methods* (5th edn.). Oxford, UK: Oxford University Press.

Chambers, R. (1994). Participatory rural appraisal (PRA): Analysis of experience. *World Development, 22*(9), 1253–1268.

Cobb, J. N., DeClerck, G., Greenberg, A., Clark, R., & McCouch, S. (2013). Next-generation phenotyping: Requirements and strategies for enhancing our understanding of genotype–phenotype relationships and its relevance to crop improvement. *Theoretical and Applied Genetics*, *126*, 867–887.

Cohen, L., Manion, L., & Morrison, K. (2018). *Research methods in education* (8th edn.). London: Routledge.

Creswell, J. W., & Creswell, J. D. (2017). *Research design: Qualitative, quantitative, and mixed methods approaches* (5th edn.). London: Sage.

Creswell, J. W., & Poth, C. N. (2016). *Qualitative inquiry and research design: Choosing among five approaches*. London: Sage.

Ellis, T. J., & Levy, Y. (2009). Towards a guide for novice researchers on research methodology: Review and proposed methods. *Issues in Informing Science & Information Technology*, *6*, 326–336.

Farid, S. (2022). Observation. In M. R. Islam, N. A. Khan, & R. Baikady (Eds), *Principles of social research methodology* (pp. 365–375). Singapore: Springer Nature.

Farley-Ripple, E. N. (2012). Research use in school district central office decision making: A case study. *Educational Management Administration & Leadership*, *40*(6), 786–806.

Foucault, M. (1972). *The archaeology of knowledge: Translated from the French by AM Sheridan Smith*. New York: Pantheon Books.

Hussein, H. (2022). Interview method. In M. R. Islam, N. A. Khan, & R. Baikady (Eds), *Principles of social research methodology* (pp. 207–219). Singapore: Springer Nature.

Islam, M. R. (2019). Designing a PhD proposal in qualitative research. In M. R. Islam (Ed.), *Social research methodology and new techniques in analysis, interpretation, and writing* (pp. 1–22). Hershey, PA: IGI Global.

Islam, M. R. (2020). Methods and framework of participatory action research for community development in Bangladesh. In R. Phillips, E. Trevan, & P. Kraeger (Eds), *Research handbook on community development* (pp. 224–243). Cheltenham: Edward Elgar Publishing.

Islam, M. R. (2021). *Poverty and inequality in Bangladesh: Regional disparity in particular reference to the Sustainable Development Goals (SDGs)*. Dhaka, BANBEIS, Ministry of Education, People's Republic of Bangladesh.

Islam, M. R. (2022a). Inquiry: A fundamental concept for scientific investigation. In M. R. Islam, N. A. Khan, & R. Baikady (Eds), *Principles of social research methodology* (pp. 3–14). Singapore: Springer Nature.

Islam, M. R. (2022b). Preparation and development of data collection instruments for social research. In M. R. Islam, N. A. Khan, & R. Baikady (Eds), *Principles of social research methodology* (pp. 449–461). Singapore: Springer Nature.

Islam, M. R. (2022c). Participatory research. In M. R. Islam, N. A. Khan, & R. Baikady (Eds), *Principles of social research methodology* (pp. 291–311). Singapore: Springer Nature.

Islam, M. R., & Hajar, A. B. S. (2013). Methodological challenges on community safe motherhood: A case study on community level health monitoring and advocacy programme Bangladesh. *Revista de Cercetare si Interventie Sociala, 42*, 101.

Islam, M. R., Cojocaru, S., Hajar, A. B. A. S., Abd Wahab, H., & Sulaiman, S. (2014). Commune and procedural level challenges and limitations in conducting social research in Malaysia: A case of disabled people. *Revista de Cercetare si Interventie Sociala, 46*, 255.

Islam, M. R., Khan, N. A., Ah, S. H., Wahab, H. A., & Hamidi, M. B. (Eds). (2021a). *Field guide for research in community settings: Tools, methods, challenges and strategies.* Cheltenham: Edward Elgar Publishing.

Islam, M. R., Khan, N. A., Ah, S. H. A. B., Abd Wahab, H., & Hamidi, M. B. (2021b). Introduction to the field guide for research in community settings: Tools, methods, challenges and strategies. In M. R. Islam, N. A. Khan, S. H. Ah, H. A. Wahab, & M. B. Hamidi (Eds), *Field guide for research in community settings: Tools, methods, challenges and strategies* (pp. 1–10). Cheltenham: Edward Elgar Publishing.

Johnstone, B. (2008). *Discourse analysis.* Malden, MA: Blackwell Publishing.

Jootun, D., McGhee, G., & Marland, G. R. (2009). Reflexivity: Promoting rigour in qualitative research. *Nursing Standard, 23*(23), 42–47.

Khan, N. A., & Abedin, S. (2022). Focus group discussion. In M. R. Islam, N. A. Khan, & R. Baikady (Eds), *Principles of social research methodology* (pp. 377–387). Singapore: Springer Nature.

Kothari, C. R. (2004). *Research methodology: Methods and techniques.* New Delhi: New Age International (P) Limited Publishers.

Lavrakas, P. J. (2008). *Encyclopedia of survey research methods.* London: Sage.

Lavrakas, P. J. (2010). Telephone surveys. In P. V. Marsden & J. D. Wright (Eds), *Handbook of survey research* (pp. 471–498). Bingely, UK: Emerald.

Mabee, S. B., Hardcastle, K. C., & Wise, D. U. (1994). A method of collecting and analyzing lineaments for regional-scale fractured-bedrock aquifer studies. *Groundwater, 32*(6), 884–894.

Mason, J. (2017). *Qualitative researching.* London: Sage.

Mohajan, H. K. (2018). Qualitative research methodology in social sciences and related subjects. *Journal of Economic Development, Environment and People, 7*(1), 23–48.

Neuman, W. L., & Robson, K. (2018). *Basics of social research: Qualitative and quantitative approaches (4th edn.).* Canada: Pearson Canada Inc.

Pannucci, C. J., & Wilkins, E. G. (2010). Identifying and avoiding bias in research. *Plastic and Reconstructive Surgery, 126*(2), 619.

Payne, G., & Payne, J. (2004). *Key concepts in social research.* London: Sage.

Russell, S. W., & Morrison, D. C. (2007). *The grant application writer's workbook: NIH version.* Burlington, MA: Jones & Bartlett Learning.

Sagor, R. (2000). *Guiding school improvement with action research.* Virginia: ASCD.

Sankar, L. V. (2022). Discourse analysis. In M. R. Islam, N. A. Khan, & R. Baikady (Eds), *Principles of social research methodology* (pp. 405–414). Singapore: Springer Nature.

Sarantakos, S. (2017). *Social research.* London: Bloomsbury Publishing.

Tenny, S., Brannan, J. M., & Brannan, G. D. (2022). Qualitative study. In *StatPearls [Internet].* Treasure Island, FL: StatPearls Publishing; 2023 Jan. Available from: <https://www.ncbi.nlm.nih.gov/books/NBK470395/>.

Wilkinson, S. (2004). Focus groups. *Doing social psychology research*, 344–376.

Yin, R. K. (2009). *Case study research: Design and methods.* New York: Sage.

Index

About the Author

DR. M. REZAUL ISLAM currently holds the position of Professor of Social Work at the Institute of Social Welfare and Research, University of Dhaka, Bangladesh. Previously, he served as a Visiting Senior Lecturer in the Department of Social Administration & Justice at the University of Malaya, Malaysia, from December 2012 to December 2016. Dr. Islam earned his Master of Social Work (MSW) and Ph.D. degrees from the University of Nottingham, England. With a career spanning over 27 years, Dr. Rezaul has been actively engaged in advancing research opportunities through international collaborations with various universities and institutions globally. He has delivered numerous lectures, public talks, and research presentations at universities worldwide. His expertise encompasses a wide range of subjects, including health and wellbeing, community development, family and child care, international and comparative social policy, climate justice, international labor migration, poverty and social inequality, and social change and globalization. Dr. Rezaul has an extensive publication record, with 130 journal articles, 50 book chapters, and 17 books published by reputable international publishers. He currently serves as a member of the International Advisory Board of the Community Development Journal (Oxford University Press) and holds editorial board positions for four journals: Asian Social Work and Policy Review (Wiley), Local Development & Society (Taylor & Francis), International Journal of Community Well-being (Springer), and SN Social Sciences (Springer). In addition to his academic role, Dr. Rezaul has collaborated extensively with international organizations such as the World Bank, UNDP, UNICEF, ILO, ADB, British Council, Plan International, Concern Universal UK, Concern Worldwide, and others. He has successfully completed approximately 30 international research projects, where he served as either a team leader or chief investigator.

www.ingramcontent.com/pod-product-compliance
Ingram Content Group UK Ltd.
Pitfield, Milton Keynes, MK11 3LW, UK
UKHW020127290926
15471UKWH00003B/332

9 781803 747187